BIOLOGY OF FRESHWATER POLLUTION

BIOLOGY OF FRESHWATER POLLUTION

C. F. Mason

Longman

London and New York

Longman Group Limited
Longman House
Burnt Mill, Harlow, Essex, UK

Published in the United States of America
by Longman Inc., New York

First published 1981

British Library Cataloguing in Publication Data

Mason, C F
Biology of freshwater pollution.
1. Freshwater biology – Great Britain
2. Water – Pollution – Great Britain
I. Title
628.1'68 QH137 80-41551

ISBN 0-582-45596-0

Printed in Singapore by
Selector Printing Co Pte Ltd

CONTENTS

ACKNOWLEDGEMENTS

We are grateful to the following publishers who very kindly gave us permission to reproduce illustrations, sometimes in a modified form, from their books and journals. We also thank the authors, who are cited in the relevant caption:

Academic Press (London) Ltd.: Fig 7.1
American Chemical Society: Fig 7.2; Table 3.2
American Public Health Association: Fig 7.6
Ann Arbor Science Publishers Inc.: Figs 3.1, 3.5
Applied Science Publishers: Fig 2.13, Table 5.2
Association of Applied Biologists: Fig 4.10
ASTM: Fig 7.7
Birkhauser Verlag: Figs 1.5, 3.11
Blackwell Scientific Publications Ltd.: Figs 1.4, 2.17, 2.20, 3.12, 4.13, 4.14, 6.3, 6.1d, 8.1, 8.2
Cambridge University Press: Figs 1.4, 4.13, 8.1, 8.2
Centre for Overseas Pest Research: Fig 7.5
Company of Biologists: Fig 2.16
Eugenics Society: Fig 2.6
Food and Agriculture Organisation of the United Nations: Figs 4.17, 4.18, 4.19
Freshwater Biological Association: Figs 3.13, 3.14
Holt, Rinehart and Winston Inc: Fig 2.15
Institute of Water Engineers and Scientists: Fig 4.3
Institute of Water Pollution Control: Fig 6.5
International Atomic Energy Agency: Fig 4.16
Journal of Fisheries Research Board of Canada: Figs 3.15, 3.16, 3.17, 3.18, 4.4
Liverpool University Press: Figs 2.7, 2.8, 2.9
Longman Group Ltd: Fig 2.12
McGraw-Hill Book Company: Fig 6.1 a, c

Munksgaard International Publishers: Fig 4.2
National Academy of Sciences: Fig 3.6
National Technical Information Service, USA: Fig 5.1
North East London Polytechnic Press: Fig 6.4
Pergamon Press Ltd: Figs 2.10, 3.2, 3.4, 3.8, 3.9, 4.5, 4.6, 4.8, 4.9, 4.11, 4.15, 4.20, 7.3, 7.4, 7.8; Tables 2.2, 3.3, 4.2, 7.1
Royal Society: Fig 4.7
Royal Society of Edinburgh: Fig 2.22
Royal Swedish Academy of Sciences: Figs 3.19, 3.20
Thunderbird Enterprises Ltd: Fig 3.3
John Wiley and Sons Inc: Table 5.1

PREFACE

The control of water pollution is important not only for amenity and public health reasons but also because clean water for domestic and industrial use is in short supply even in comparatively wet countries such as the British Isles. The water shortage that developed in England and Wales during the hot, dry summer of 1976, causing severe problems to agricultural, industrial and domestic users, provided ample proof of this.

It is the biological effects of pollution, including those on man, which are of greatest importance, so it would seem natural that biologists should be involved in pollution control. Until recently, however, the management of water resources has been largely the domain of the engineer. The great increase in the number of biologists employed in the water industry over the last decade is evidence that biological expertise has much to offer in the fields of pollution detection and assessment, and in the management of the water cycle as a whole.

Since the publication of H. B. N. Hynes' classic *The biology of polluted waters* in 1960 there has been a tremendous increase in knowledge concerning pollution, particularly in the fields of eutrophication, toxic pollution and pollution assessment. The present work provides an introduction to the biological effects of water pollution and to ways of detecting, describing and quantifying these effects in the field and in the laboratory. It is intended not as a review, but an overview, which has had, of necessity, to be highly selective. At all stages, however, I have referred to modern reviews so that the reader can follow up, in depth, areas of particular interest to him. Some knowledge of the fundamentals of freshwater biology and of the factors which influence the distribution of animals and plants has been assumed. The book should enable a student to see how the academic content of much of the curriculum is used daily by biologists in an applied manner and it should provide enough practical information for interested students to carry out pollution assessments of their own.

I am extremely grateful to Drs J. W. Hargreaves, S. M. Macdonald and R. S. Wilson for reading the whole of the text in draft, and to Professor A. Macdonald, Dr R. D. Roberts and Dr S. P. Long for comments on Chapters 2, 4 and 8 respectively. I am also indebted to Miss Vivien Amos for typing the various drafts of the manuscript.

C. F. Mason
Wivenhoe, July 1980

Chapter 1

INTRODUCTION

What is pollution?

There are a number of definitions of pollution in current usage. A recent dictionary of life sciences (Martin, 1976) defined pollution as 'the presence in the environment of significant amounts of unnatural substances or abnormally high concentrations of natural constituents at a level that causes undesirable effects, such as bronchial irritation, corrosion or ecological change'. Such a definition is probably too broad to be useful. Streams flowing through deciduous woodlands receive large autumnal inputs of leaves which may de-oxygenate the water and result in an impoverishment of the fauna. To the angler, wishing to catch trout, the leaves are pollutants, though the ecologist would consider de-oxygenation as a normal seasonal feature in the dynamics of such streams. The angler would be similarly distraught if he fished the headwaters of streams flowing from granite, in which the input of nutrients and energy is too low to support more than an impoverished fauna.

To avoid the consideration of naturally stressed environments, definitions of pollution are often restricted to include only the effects of substances or energy released by man himself on his resources (Edwards *et al*, 1975). The definition followed in the present book is that of Holdgate (1979):

> The introduction by man into the environment of substances or energy liable to cause hazards to human health, harm to living resources and ecological systems, damage to structure or amenity, or interference with legitimate uses of the environment.

Why need we be concerned about pollution?

From the definition above, it would seem obvious that pollution is important because man's resources are being damaged. Pollution emanates from man himself and from his activities so he should be able to control pollution. However, pollution control is extremely costly and the benefit in resource terms may be far outweighed by the cost of control. Furthermore individual (or nation) 'A' might, through pollution, be damaging the resources of individual (or nation) 'B' rather than his own and he may be unwilling to reduce his profits to benefit his neighbours. Stringently applied laws, or a high degree of altruism, are required to control pollution.

The management of water resources must be done in such a way that the water provides the capacity for sustaining life for the existing population and social activities of different scales and provides the potential for further expansion and growth, thus enabling an improvement of the social wellbeing by an upgrading of the quality of life (Lindh, 1979).

We can look at the water resources of England and Wales as a specific example of why pollution control is important. A generalized diagram of the water cycle is given in Fig. 1.1.

The population of England and Wales in 1975 was estimated at 49.2 m. people, occupying a land area of 151 139 km^2 and giving an average density of 325 people per square kilometre. The actual distribution of population of high density is shown in Fig. 1.2.

The average consumption of water in England and Wales in 1975 was equivalent to 307 litres per person per day, which amounted to 5500×10^6 m^3 per annum. The residual rainfall (Fig. 1.3) in 1975 was 360 nm, which amounted to a total of $54\,410 \times 10^6$ m^3 of water. The residual rainfall is the difference between precipitation and actual evaporation (including transpiration) and represents the total amount of water available to replenish rivers, aquifers, lakes and reservoirs; in other words it is the total quantity of water theoretically available for use. Not all of this is actually available; for instance, rain falling in coastal areas may go directly into the sea. Nevertheless, if we divide the residual rainfall by the actual consumption, the theoretical water resources in England and Wales in 1975 were some ten times more than the actual demand. If one includes direct abstractions by industry and agriculture, the total resources are some five times greater than demand. These figures do not take into account the considerable re-use of water which increases the discrepancy between resources and demand.

However, the demand for water is greatest during the summer, whilst most rain falls during the winter. Also the rainfall is not evenly distributed over England and Wales. If the distribution of residual rainfall (Fig. 1.3) is compared with the regions of high population

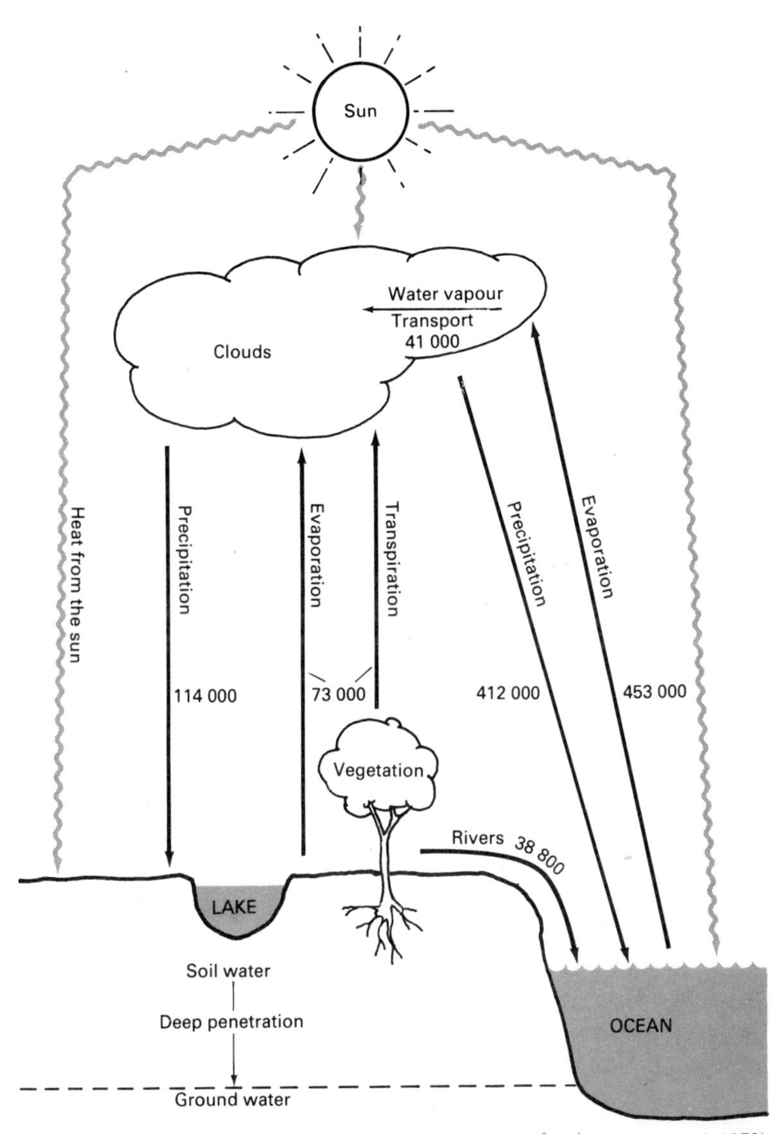

Fig. 1.1 The hydrological cycle. Global flows of water in $km^3\ y^{-1}$ (from Lvovich 1973).

(Fig. 1.2) we can see that a number of regions of high population and industry, for instance the Midlands, London and the south-east, are in areas of low rainfall. Furthermore, arable farming, with its requirements for irrigation, is also concentrated in the drier areas of the country. Annual rainfall, of course, also varies and a ten-year

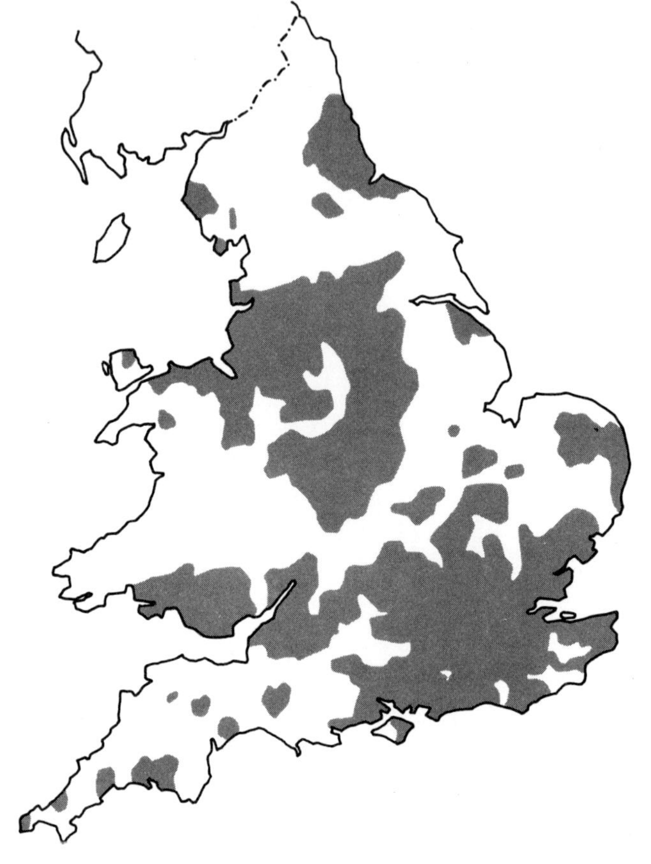

Fig. 1.2. The areas of high population density (shown black) in England and Wales.

periodicity has been demonstrated for the rainfall of England and Wales (Rodda *et al*, 1978).

This discrepancy between water availability and water use means that the actual resources may fall at times below demand, as happened in the exceptional drought year of 1976, when considerable restrictions on water use became necessary. There are similar problems of supply and demand in many other areas of the world, e.g. the Potomac River in Washington D.C. can no longer safely supply the population demands for water in dry years (Withers, 1978). Evapotranspiration exceeds precipitation in large areas of the world, including many developing countries, where rational water management is of paramount importance (Kalinin and Shiklomanov, 1974).

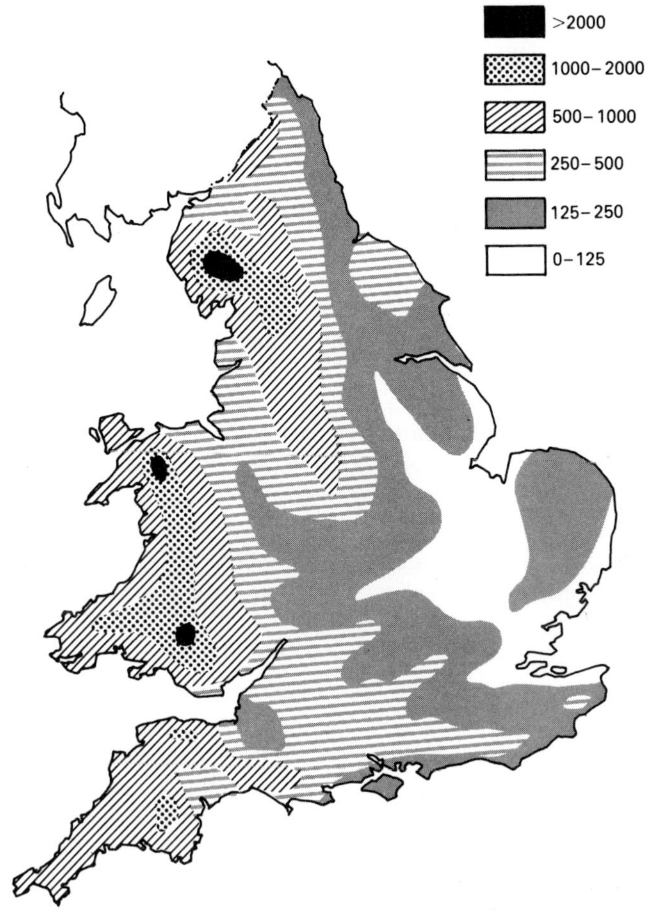

Fig. 1.3 The residual rainfall (mm) in England and Wales.

The disparity between areas of high rainfall and areas where water is needed result in large quantities of water in England and Wales being abstracted for the public water suppply from the lowland reaches of rivers and there is also considerable direct abstraction by industry and agriculture. *Such water must be of an acceptable quality*. Abstractions from the lower reaches of a river may be sustained by a controlled discharge from a reservoir usually situated in the head-waters where rainfall is heavier, this being an economical way of transporting water to where it is needed. Water may also be transferred from one catchment to another which is short of water. Many rivers are being used in this way as aqueducts.

Domestic, industrial and agricultural users produce large quantities of waste products and waterways provide a cheap and effective way of disposing of many of these. During dry weather, the flow of some rivers consists almost entirely of effluents. The effluents of some towns become the water supplies of other towns downstream. For this reason alone it is important that the effluent discharged into a watercourse is of high quality and the degree of pollution is such that the self-purifying capacity of the river (see p.34) is not overloaded.

In addition to providing a source of water and a sink for effluents, freshwaters have an important amenity role, including such activities as boating, angling and wildlife studies. Some of these pastimes require water of a very high quality. In 1970, some 2.79 m. people went fishing in England and Wales, making angling by far the largest participating sport in the country (Natural Environment Research Council, 1971). The service industries for these pastimes are locally very valuable. There are also lucrative commercial fisheries for salmon, migratory trout and eels. It is also very desirable that the natural communities of animals and plants in freshwaters be maintained and in England and Wales the Regional Water Authorities have a statutory requirement to consider the biota of freshwaters in any management scheme.

Finally, the resources of the seas are vast and most of the pollutants travelling down rivers will eventually end up in the sea. Animals such as arctic seals and antarctic penguins are already loaded with pollutants and we should not be complacent that the immense quantities of water in the oceans can absorb pollution indefinitely without effect.

There are, therefore, a number of reasons why the control of pollution is important. We do not, of course, have equal concern for all the components of our environment and this is especially so when the enormous costs of pollution abatement are taken into account. Holdgate (1979) has ranked target organs for pollutants in order of decreasing concern to man:

Man → Domestic livestock → Crops and structures → Most wildlife and amenity → Pests and disease vectors

Man's prime concern is to reduce the risk of pollution affecting his own health, whilst at the other extreme, if pollution kills organisms which are pests to man then it might be considered positively beneficial. Some groups of organisms may be of fairly low priority in terms of man's, or at least politicians', concern, but they may require water of an especially high quality. We can list a number of potential uses of water in terms of decreasing water quality requirements (Poels *et al*, 1978):

nature reserves → recreation → fisheries → water for potable supply → watering cattle → irrigation → processing water → cooling water → shipping.

Thus nature reserves may need completely unpolluted water (though this is not necessarily so, it depends rather on what is being conserved), whereas a grossly polluted water will suffice for shipping. Note that because of the efficiency of water treatment processes water for potable supply need not be of the highest quality.

It would obviously not be economically feasible to clean all waters to such an extent that they would make pristine nature reserves and economic considerations may make it unrealistic to improve the quality of some waters for recreation and fisheries. Increasingly the concept of maintaining water quality at a standard relating to the use to which that water is put is being adopted. Thus a classification system for water uses is constructed and for these uses water quality criteria are formulated (see p.192).

What are pollutants?

In terms of the definition given at the beginning of this chapter almost anything produced by man can be considered at some time to be a pollutant. Indeed, to the farmer whose land is about to be lost under a new reservoir scheme, pure water is itself a pollutant in almost every sense of the definition given. Substances which are essential to life (e.g. copper, zinc) can be highly toxic when present in large amounts.

Some 1500 substances have been listed as pollutants in freshwater ecosystems and a generalized list is given in Table 1.1. Some of the categories are not necessarily mutually exclusive, e.g. domestic sewage may contain, in addition to oxidizable material, detergents, nutrients, metals, pathogens and a variety of other compounds.

Whether or not a compound will exert an effect on an organism or a community will depend on the concentration of that compound and the time of exposure to the compound (i.e. the dose). The effect of a pollutant on a target organism may be either acute or chronic (Saunders, 1976). Acute effects occur rapidly, are clearly defined, often fatal and rarely reversible. Chronic effects develop after long exposure to low doses or long after exposure and may ultimately

Table 1.1. Categories of pollutants found in freshwater.

Acids and alkalis.
Anions (e.g. sulphide, sulphite, cyanide).
Detergents.
Domestic sewage and farm manures.
Food processing wastes (including processes taking place on the farm).
Gases (e.g. chlorine, ammonia).
Heat.
Metals (e.g. cadmium, zinc, lead).
Nutrients (especially phosphates and nitrates).
Oil and oil dispersants.
Organic toxic wastes (e.g. formaldehydes, phenols).
Pathogens.
Pesticides.
Polychlorinated biphenyls.
Radionuclides.

cause death. Sub-lethal doses result in the impairment of the physiological or behavioural processes of the organism (e.g. it may grow poorly, or fail to reproduce). Its overall fitness is reduced.

At the community or ecosystem level it is unlikely that pollution will cause irreversible effects, except possibly in the case of radioactive pollution. The effects of pollution are recorded in the loss of some species, with possibly a gain in others, generally a reduction in diversity, but not necessarily numbers of individual species, and a change in the balance of such processes as predation, competition and materials cycling.

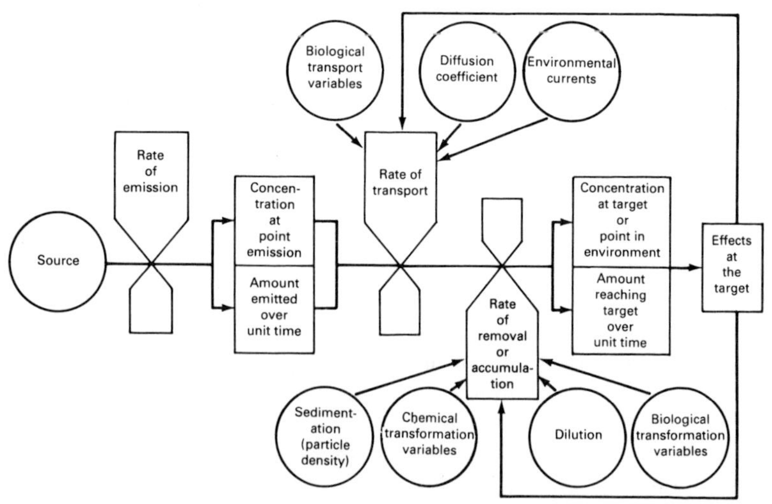

Fig. 1.4. A generalized pollutant pathway (from Holdgate 1979).

The generalized pathway of a pollutant from source to target is shown in Fig. 1.4. There are three important rate processes in the pathway: the rate of emission from the source of pollution; the rate of transport through the ecological system; and the rate of removal or accumulation of the pollutant in the pathway. The rate of transport will depend on the diffusion rate of the pollutant and on a variety of environmental factors as well as properties of transport within organisms in the pathway. The rate of removal or accumulation will depend on rates of dilution or sedimentation and on chemical and biological transformations. These determine the dose reaching the target organisms. Processes within the target will either transport the pollutant to where it exerts an effect or will excrete the pollutant. We can see how this basic pathway can be repeated along a food chain. Also, with slight terminological changes, Fig. 1.4. can illustrate the basic pathway from entry point to site of action within a target organism.

Monitoring pollution

Holdgate (1979) has pointed out that to effectively manage an environment receiving polluting substances we need information concerning:

1. The substances entering the environment and their quantities, sources and distribution.
2. The effects of these substances within the environment.
3. Trends in concentration and effect, and the causes of these changes.
4. How far these inputs, concentrations, effects and trends can be modified and by what means at what cost.

The first stage in this management is to carry out a *survey*, which is a programme of measurements that defines a pattern of variation of a parameter in space. As an example, we may be concerned about the output of zinc into a river in an effluent from a rubber-processing factory. Our initial survey may involve measuring the levels of zinc in the river sediment at a number of stations downstream from the factory, together with sampling the fauna and flora at these stations. The survey will only inform us of the situation at one point in time.

The next stage will be *surveillance* and *research*, which will enable us to learn more about a problem before any policy decisions are made. Surveillance is defined as the repeated measurement of a variable in order that a trend may be detected. In our example we may measure the level of zinc in the sediments at three-monthly intervals to see how sedimentary loadings vary. Similarly the animals and plants will be sampled to see if the original observations are repeated. The research function will be to examine the pollution process in more detail, using experimental and analytical techniques. For instance the survival of fish in concentrations of zinc flowing from the rubber processing factory could be studied in the laboratory, or experimental streams might be used to study whole communities. Furthermore, the tolerance of organisms to concentrations of zinc lower than that in the effluent might be studied as a guide to fixing a standard for the zinc level.

From the surveillance and research programme, and taking into account economic considerations, a policy for managing the pollution might be decided. For example the level of zinc in the effluent might be reduced by 75 per cent by installing a treatment plant in the factory. It will be necessary to see that this reduction has the desired effect, an improvement in the state of the receiving river, and also to ensure that the effluent's quality is maintained. These observations on performance in relation to standards are known as *monitoring*.

It should be noted that these terms are used rather loosely and the boundaries between them are indistinct. Furthermore there is not necessarily an ordered sequence in the procedure. The monitoring programme set up might produce useful research data which can be used to redefine the policy, which may then require modifications in the monitoring strategy.

The strategy for examining pollution outlined above involves both the experimental and observational approach (Fig. 1.5). The experimental approach involves the simulation of the pollutant's behaviour and its action on resources in experimental systems, whereas the observational approach examines the distribution of pollutants

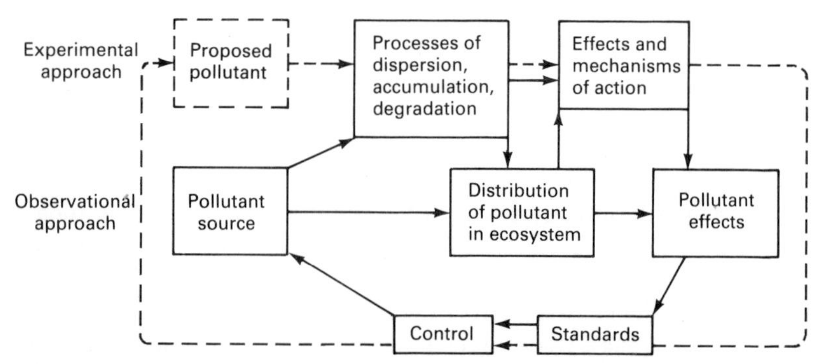

Fig. 1.5. A flow diagram of pollution and its control (from Edwards 1975).

and their pattern of effects on natural resources (Edwards, 1975). The integration of these two approaches is essential if a management policy for a particular pollutant or pollutional source is to be effective. Both approaches have their disadvantages. The experimental simulation does not take into account the complexity of the normal pollutional situation in which a variety of factors influence the way a pollutant affects its target. This very complexity, however, makes the interpretation of observational data exceedingly difficult.

Why biological surveillance?

When carrying out a pollution surveillance programme, why do we not merely measure the levels of pollution at key points in the transfer pathway, especially if for economic reasons, we are only interested in man and his food supplies? Pollutant levels can be measured at the sites of discharge and the sites of abstraction from the watercourse for potable supply, irrigation etc., and we need not concern ourselves with problems in the river. Moreover, the concentrations of chemicals can be measured accurately and repeatedly, using standard methods (American Public Health Association, 1971; Department of the Environment, 1972; with many of the standard tests for use in student projects in Best and Ross, 1977). Biological material is notoriously difficult to sample and inherently variable. However, there are considerable advantages in biological surveillance.

Animal and plant communities respond to intermittent pollution which may be missed in a chemical sampling programme. For example, a river may be sampled at station 'A' every Tuesday and analysed, back in the laboratory, for ten chemical determinands, one of which is zinc. Such regular programmes may be necessary for logistical reasons where a large number of sites are under surveillance. On Wednesday morning each fortnight a factory immediately upstream may be discharging an effluent containing zinc. By the following Tuesday this will have disappeared downstream and the chemist will not detect zinc. However, a biologist, sampling at monthly intervals alongside the chemist, will record an unexpected depression in the diversity of the biological community, as some species may be eliminated and many individuals killed by the zinc. Some of the species missing may be known to be especially sensitive to zinc, i.e. are acting as *indicators*. As replacements of organisms at station 'A' will have to be by immigration and reproduction the polluting incident will be apparent for several weeks or even months after the event.

The chemist may get round the problem of periodic sampling by installing an automatic analyser at the station, but these are very expensive, they deal with only a few determinands and they are liable to failure under the often rigorous and unpredictable conditions in the field.

This brings us to the next advantage of biological surveillance – biological communities may respond to new or unsuspected pollutants in the environment. It would obviously be uneconomic and impracticable to regularly determine concentrations of the 1500 or so known pollutants. The water industry at present regularly tests for about thirty determinands. However, if a change in a biological community, or in members of that community, is detected and gives cause for concern, then a detailed screening for pollutants, and indeed of chemicals hitherto not considered as pollutants, can be made. One example of this was a large and unexplained mortality of seabirds in the Irish Sea in September 1969. An analysis of these birds suggested that high levels of polychlorinated biphenyls were involved in the mortality and provided an early indication that PCBs, released in effluents from some chemical industries, were contaminating inshore waters (Natural Environmental Research Council, 1977). Prior to this PCBs had received little consideration.

The example of PCBs in seabirds demonstrates another advantage in biological monitoring, namely that some chemicals are accumulated in the bodies of some organisms and the levels can reflect the environmental pollution levels. These accumulations may be built up over a long period of time, while at any particular point in time the pollutant may be present at too low a level in the water to be detected without the concentration of large quantities of water.

In this way many organisms may be used in bio-assays in surveillance programmes. Not only may this involve concentrating pollutants but it may also include growth responses (e.g. of algae in nutrient rich water) or mortality (e.g. of fishes placed in effluents).

It is worth remembering that the uses to which man puts water (drinking water for himself and his livestock, irrigation, food processing etc.) often have biological implications. The effects which are perceived in natural communities might be considered as an early warning system for the potential effects of pollutants on man.

The complexity of pollution

The following chapters will show the general characteristics and effects of various types of pollutants. However, it must be stressed that it is only rarely that a single pollutant is present in a watercourse. Normally an effluent will consist of a variety of potentially harmful substances and most watercourses will receive a number of effluent discharges and so the effects of these will often be difficult or impossible to disentangle.

Pollutants occurring together may act completely independently on a target and the one exerting the greatest effect would then be the most important. One would not, for example, worry unduly about high levels of zinc in an effluent if the oxygen demand was so high that all life in the receiving stream was suffocated, but if the organic loading in the effluent was reduced such that the stream could support life, the concentration of zinc might then become important. The effects of pollutants might also be additive, or antagonistic (in which the combined effect on the target organisms is less than predicted by each pollutant's effect when alone) or synergistic (when the combined effect is greater than predicted from their effects when alone). These interactions will become apparent in later chapters.

Chapter 2

ORGANIC POLLUTION

Introduction

Organic pollution results when large quantities of organic compounds, which can act as substrates for micro-organisms, are released into watercourses. During the decomposition process the dissolved oxygen in the receiving water may be utilized at a greater rate than it can be replenished, causing oxygen depletion, which has severe consequences for the stream biota. Organic effluents also frequently contain large quantities of suspended solids, which reduce the light availability to autotrophs and, on settling out, alter the characteristics of the river bed, rendering it an unsuitable habitat for many organisms. Ammonia is also often present and this adds to toxicity.

A simple measure of the potential of biologically oxidizable matter for de-oxygenating water is given by the biochemical oxygen demand (BOD). The BOD is obtained in the laboratory by incubating a sample of water for five days at 20 °C and determining the oxygen utilized. Effluents with high BODs can cause severe problems in receiving watercourses.

Organic pollutants consist of proteins, carbohydrates, fats and nucleic acids in a multiplicity of combinations. Dugan (1972) lists almost ninety organic compounds or groups of compounds excreted by man and potentially finding their way into rivers. Organic wastes originating from man and his animals may also be rich in pathogenic organisms. Additionally pharmaceuticals, such as antibiotics and hormonal contraceptives, administered to man and his animals, have been suggested as possible pollutants. At least as far as hormonal contraceptives are concerned, levels recorded in drinking water are considered too low to present a hazard (Rathner and Sonneborn, 1979).

Origins of organic pollutants

Organic pollutants originate from domestic sewage (raw or treated), urban run-off, industrial (trade) effluents and farm wastes.

Sewage effluent is the greatest source of organic materials to freshwaters. In England and Wales in 1975 there were 4129 discharges releasing treated sewage to rivers and canals and a further 414 discharges of crude sewage, 95 per cent of these latter being to the lower, tidal reaches of rivers (Department of the Environment, 1978). The total daily volume of sewage effluent in 1975 amounted to 11.2×10^6 m^3, the majority of which was of domestic sewage, but with an average of some 17 per cent of industrial effluent, as well as urban run-off. In addition, about 1×10^6 m^3 per day of untreated sewage in England and Wales, from outfalls serving 6–10 per cent of the total population, is discharged directly to the sea, where it is assumed that the capacity for purifying biodegradable matter is almost unlimited (Department of the Environment, 1973). Sewage treatment processes in England and Wales remove 77 per cent of the overall polluting load, measured in terms of BOD, and this compares favourably with other European countries. Some 95 per cent of households in England and Wales are connected to public sewers, compared with 75 per cent in the United States, 80 per cent in Federal Germany and 88 per cent in the Netherlands (National Water Council, 1978). Of course, in the less developed countries of the world, where the capital expenditure required for providing sewage treatment facilities is not available, the release of crude sewage into watercourses is still a very major problem. The present-day cost of providing a sewage treatment works capable of producing a satisfactory effluent is enormous, in Britain somewhere in the region of £10 m. for a population of 100 000.

In urban areas, the run-off from houses, factories and roads can result in severe pollution, especially in storm conditions after periods of dry weather. This urban run-off may be routed through the sewage works, or it may be separately sewered and flow directly into rivers and streams. In the former case sewage treatment works may be severely overloaded during storm conditions and the effluent discharged may be of a much lower quality than under normal operating conditions, though the dilution is greater when the receiving river is in flood. Where urban run-off drains directly into rivers, pollution can be severe because drainage from the hard surfaces is so rapid that it may reach a river which has not yet increased above dry weather flow, so that the dilution will be minimal.

With urban run-off there is often a first flush of highly polluting water and much of this may come from the catchpits of roadside drains, which have thick bacterial scums and are usually anoxic. The constituents of urban run-off are obviously very variable. The Water Research Centre (1977) estimated that 17 $g\ m^{-2}\ y^{-1}$ of dog faeces alone will form part of urban run-off and the effluent will also be rich in suspended material, heavy metals and, seasonally, chloride from road-salting operations. The quality of the run-off is also highly variable, but BODs as high as 7700 $mg\ l^{-1}$ have been recorded.

Industrial effluents are a further source of organic pollution. These may be routed via the sewage treatment works or they may be released, with or without treatment, directly into a waterway. Amongst the industries producing effluents containing substantial amounts of organic wastes are the food processing and brewing industries, dairies, abattoirs and tanneries, textile and paper making factories. Excluding minewater discharges and those industries which released effluents to sewers, the Department of the Environment (1978) recorded 2959 discharges to rivers and canals in England and Wales in 1975 and these released $68.7 \times 10^6\ m^3$ of effluent daily, though some 87 per cent of this volume consisted of cooling water alone, which should contain only insignificant concentrations of organic pollutants. The volumes of process water released per day by some selected industries is given in Table 2.1.

Table 2.1. The volumes of discharges of process water from selected industries to rivers and canals in England and Wales in 1975 (from Department of the Environment, 1978).

Industry	*Volume of process water* $m^3\ d^{-1}$	*Industry*	*Volume of process water* $m^3\ d^{-1}$
Brewing	10 680	Metal smelting and refining	30 490
Chemical	385 550	Paper and board making	446 570
Engineering	51 210	Plating and metal finishing	24 360
Food processing	279 510	Petroleum refining	126 840
Gas and coke making	60 120	Textile manufacture	94 110
Glue and gelatin manufacture	22 110	Trade effluent from quarries and mining	482 670
Iron and steel making	185 880	Disposal tip drainage	18 620

Many trade effluents can be effectively treated by mixing with domestic sewage, but some can inhibit the microbiological activity in the treatment works, resulting in a poor quality effluent, and these must be treated on the industrial site.

Farm effluents have become an increasing pollution problem, particularly with the intensification of livestock production in recent years, and this is dealt with in more detail in Chapter 3. The liquor from badly sited dumps of farmyard manure can cause serious pollution, especially when it is washed into drainage ditches after periods of rain. Such sources of pollution can be very difficult to trace. The washing down of intensive rearing units and milking parlours can also result in highly polluting wastes entering streams.

Pollution is also caused by some of the more intensive and mechanized ways of dealing with vegetable crops. For instance, the effluent released from silos, in which grass and other fodder crops are partially fermented, can be 220 times as strong, in terms of BOD, as settled sewage (Klein, 1962). The mechanical vining of peas and especially the drainage from pea haulm silage can cause severe, though seasonal, problems. One such effluent had a BOD of 15 000 mg l^{-1} (Klein, 1962).

The origins of effluents and their composition are thus extremely diverse and we can expect a similar diversity in the effects they have on receiving waters.

Pathogens

The faecal contamination of water can introduce a variety of pathogens into waterways. The most common of these have been listed by Geldreich (1972) as strains of *Salmonella*, *Shigella*, *Leptospira*, enteropathogenic *Escherichia coli*, *Francisella*, *Vibrio*, *Mycobacterium*, human enteric viruses, cysts of *Entamoeba hystolytica*, and parasitic worms.

Salmonella can cause acute gastro-enteritis (i.e. food poisoning), with diarrhoea, fever and vomiting, and typhoid, caused by *S. typhi*, is an example. Several hundred serotypes of *Salmonella* are known to be pathogenic to man and there are many more which infect livestock. Cross-infection between man and animals can occur via water pollution. There appears to be an increase in the spread of *Salmonella* bacteria and this has been related to modern living conditions, such as mass food production and communal feeding (Windle-Taylor, 1978). Salmonellae in rivers can survive for some distance downstream of their source, e.g. they have been recorded 117 km below the nearest sources of animal pollution in the Red River, North Dakota (Geldreich, 1972).

Shigella is the most commonly identified cause of acute diarrhoea in the United States (Geldreich, 1972) and a number of epidemics have been traced to poor quality drinking water. The isolation of *Shigella* from polluted water is difficult and the survival of the bacterium appears to be low.

Leptospirosis results in an acute infection of the kidneys, liver and central nervous system and is caused by a group of spiral-shaped, motile bacteria, of which more than 100 serotypes are known. The primary hosts are rodents, in which the bacteria are carried in the kidneys, and man may become infected by drinking, or by wading or swimming in contaminated water. Most cases occur from infection through skin abrasions or from accidentally falling into contaminated water. Weil's disease is a particularly serious form of leptospirosis and all workers likely to come into contact with sewage in the United Kingdom now carry special cards, warning doctors of their potential exposure to this disease, should they become ill.

Gastro-enteritis and diarrhoea, especially in young children, may be caused by various serotypes of *Escherichia coli*. Most urinary infections of adults are also caused by pathogenic *E. coli* (Geldreich, 1972), though urinary infection is usually by retrograde spread of *E. coli* from the person's own intestinal flora, rather than from water supplies. *Campylobacter* species have recently been recognized as a cause of large outbreaks of intestinal infections. Unpasteurized milk is probably the main source, but water-borne sources are considered important in the USA.

Tularemic epidemics, caused by *Francisella tularensis*, result in chills and fever and a general weakness which may last for a number of weeks (Geldreich, 1972). The pathogen enters man through skin abrasions or the mucus membranes. The disease reaches epizooitic proportions in wild rodents and rabbits and infects man via water contaminated by the animals' urine, faeces or corpses.

Vibrio cholerae causes cholera, an acute intestinal disease which may result in death within a few hours of onset. The bacterium lives on the gut wall and produces an exotoxin consisting of two parts. The B portion binds onto cell surface receptors in the gut wall, while the A portion acts as a hormonal mimic, which causes the cell to greatly increase its output of water, sodium bicarbonate and potassium. This fluid then leaves the body as 'rice-water' diarrhoea, which helps to disseminate the bacteria, but results in severe dehydration of the human host, with attendant breakdowns in the circulation of the blood. Cholera is largely under control in countries where widespread sewage treatment is practised, but it is still rife in many parts of the world and outbreaks occur when disasters happen, e.g. famine, earthquakes and floods. The factors which determine the effect of different water supplies on the occurrence of outbreaks of cholera are discussed by Jusatz (1977). Some 150 000 cases from 36 countries were reported in 1970 (Windle-Taylor, 1978). Honda and Finkelstein (1979) have recently isolated a strain of *V. cholerae* which produces only the B portion of the toxin, so that it may be possible eventually to take orally a living vaccine which will produce immunity without stimulating fluid production. This could prevent isolated cases of cholera from developing into epidemics.

Mycobacterium is responsible for tuberculosis. Transmission by water appears to be uncommon, but this may in part be due to the long time between infection and the appearance of symptoms, which makes the origins of the disease difficult to trace in any particular incident. The tubercle bacilli are able to survive in water for several weeks.

More than 100 different kinds of enteroviruses are known and they are present in sewage and receiving waters. Infectious hepatitis (jaundice) is endemic in many countries and, of the two main kinds, hepatitis A is transmitted by contaminated water. Like cholera, it thrives during disasters, such as floods, or in refugee camps. Poliomyelitis may occasionally be carried by water, but most transmission is by personal contact. Enteritis can be caused by coxsackie and ECHO viruses and outbreaks may develop very rapidly if water supplies become contaminated with untreated sewage. Viruses appear to survive in water for much longer than faecal bacteria and there is increasing concern over their occurrence and control in sewage effluents and in water supplies.

Amoebic dysentery, caused by the parasitic protozoan *Entamoeba histolytica*, is endemic in the tropics and subtropics and the faecal contamination of drinking waters is a major source of transmission. Other diseases are due to protozoa, including the rare, but fatal, amoebic meningoencephalitis, caused by a pathogenic strain of *Naegleria fowleri* which is contracted while swimming in the warm waters of small lakes, indoor swimming pools or a polluted estuary (Geldreich, 1972).

The final group of pathogens listed by Geldreich (1972) is the parasitic worms. These include the phyla Nematoda (roundworms) and Platyhelminthes (tapeworms and flukes). The ova of these parasites are passed out in the faeces and urine, are often resistant to sewage treatment processes and a tapeworm can lay up to 1 m. eggs in a day. The majority of life cycles involve an intermediate host. In the case of the beef tapeworm, *Taenia saginata*, cattle, the intermediate host, may become infected by grazing on pastures sprayed with sewage sludge or by drinking water contaminated with sewage.

Schistosomiasis (bilharzia) is of particular importance in that 200 m. people, mainly in the tropics, have the disease, which is caused by blood flukes of the genus *Schistosoma*. It is a debilitating disease which, though not often fatal, causes general weakness in the sufferer and results in an annual economic loss of some 560 m. dollars (Benarde, 1970). Eggs of the flukes pass out with the faeces or urine of man and, if they reach freshwater, they develop into miracidium larvae which infect snails. Cercariae develop in the snails and, on leaving the host, they penetrate the skin of humans wading in the water. They mature in the bloodstream, showing a particular preference for the portal vein, carrying nutrient-laden blood from the intestine to the liver.

The incidence of bilharzia appears to be influenced by changes in land-use. With the filling of Lake Nasser behind the Aswan Dam in Egypt the surrounding land is now irrigated by canals and ditches rather than by seasonal flooding. This has led to the build-up of large populations of snail hosts, which now have a permanent rather than a seasonal habitat, and the incidence of bilharzia has greatly increased (Sterling, 1971; Hammerton, 1972). Similarly forest clearance and agricultural development in Cameroun has led to the appearance of *Schistosoma haematobium*, because the new conditions have allowed its snail host *Bulinus rohlfsi* to spread (Southgate *et al*, 1976). *Schistosoma haematobium* is replacing the indigenous *S. intercalatum* through introgressive hybridization. *Schistosoma haematobium* has a number of advantages over *S. intercalatum* in the Cameroun situation and may result in an increase in the incidence of schistosomiasis.

Fascinating general accounts of the ecology of pathogens and parasites are given by Burnet and White (1972) and Baer (1971). To avoid these many water-borne diseases it is obviously essential that as many pathogens as possible are removed from sewage before it enters a watercourse.

Sewage treatment

Historical

Until the last 200 years or so the deterioration of watercourses due to organic pollution was not a serious problem because a relatively small human population lived in scattered communities and the wastes dumped into rivers could be coped with by the natural self-purification properties. Extrusive, pathogenic organisms, by contrast, were a severe problem though the link between disease and pathogens transmitted by faecal contamination had not of course been made.

Water pollution became a severe problem with the industrialization of nations, coupled with the rapid acceleration in population growth. Industrialization led to urbanization, with people leaving the land to work in the new factories. Domestic wastes from the rapidly expanding towns and wastes from industrial processes were all poured untreated into the nearest river.

The historical pollution of the River Thames has recently been eloquently described by Doxat (1977). Though pollution in the Thames at London was recorded as early as the thirteenth century, it was not until the eighteenth century that problems became acute. The population of London doubled between 1700 and 1820 to 1 250 000. Sewage was collected in individual or communal cesspits, which were periodically emptied to fertilize the surrounding land. In 1843 main sewers were laid and 200 000 cesspits were abolished. The sewage drained directly into the Thames, causing gross pollution and epidemics of cholera, whilst noxious odours rising from the river periodically disrupted the work of Parliament and the Law Courts.

In 1865 a scheme designed by Sir Joseph Bazalgette diverted the sewage, via three main sewers, to outfalls situated 10 miles below London Bridge, where it was discharged untreated on the ebb tide. The river through London showed some improvement, though sewage and industrial effluents from many other discharges continued to add to the pollution. The situation around the outfalls downstream of the capital was atrocious.

During the latter part of the nineteenth century a Rivers Pollution Committee recommended various treatment processes and in 1882 a Royal Commission on Metropolitan Sewage Disposal was set up and eventually concluded that the suspended solids in sewage should be separated from the liquid before it was discharged into the river. At this time Mr W. J. Dibdin put before the Commission the idea that sewage could be treated biologically, though this was not taken further by the Commission. The solids in sewage were separated at the outfalls from 1889 and the resulting sludge was dumped at sea.

Dibdin's ideas on bacteriological treatment of sewage was researched further through the 1890s and the first treatment plant was put into commission in 1914–though in Manchester, not London.

Three experimental plants for biological treatment were installed down-stream of London in 1920 and a major plant was erected in 1928, with greatly expanded facilities provided since then. A marked improvement in the quality of the water in the River Thames has become apparent during the 1960s and a diverse flora and fauna, absent for many years, has begun to colonize the river.

Similar stories could be told of cities throughout the world.

The basic treatment process

Price (1970) has listed three objectives in the treatment of sewage; to convert sewage into suitable end products, that is an effluent which can be satisfactorily discharged into the local watercourse and a sludge which can be readily disposed; to carry this out without nuisance or offence; and to do so economically and efficiently.

There are four stages in the treatment of sewage but, depending on the quality of the effluent required, not all the stages may be utilized:

(a) *preliminary treatment*, which involves screening for large objects, maceration and the removal of grit, together with the separation of storm flows;
(b) *primary treatment* (*sedimentation*), where the suspended solids are separated out as sludge;
(c) *secondary* (*biological*) *treatment*, where dissolved and colloidal organics are oxidized in the presence of micro-organisms;
(d) *tertiary treatment*, which is used when a very high quality effluent is required. It may involve the removal of further BOD, bacteria, suspended solids, specific toxic compounds or nutrients.

The amount of treatment which is provided will depend to some extent on the amount of dilution which is available in the receiving water and on the quality objectives for that water. Preliminary treatment only may be given for effluents being discharged out to sea, whereas tertiary treatment may be required if water is abstracted for potable supply downstream of a discharge. In Britain the minimum requirement for an effluent, which can be achieved with secondary treatment, is the Royal Commission Standard, allowing no more than 30 mg l^{-1} of suspended solids and 20 mg l^{-1} BOD (a 30:20 effluent). To achieve these standards, the Royal Commission envisaged that the effluent would be diluted with eight volumes of clean river water, having a BOD of 2 mg l^{-1}. Such a dilution may not always be available so that a more stringent standard than 30:20 may be required for the effluent.

A modern sewage treatment work is illustrated in Figs. 2.1 and 2.2. The Crossness works serves a population of 1.7 m. people and treats an average of 580 000 m^3 of sewage per day.

Preliminary treatment

The sewage is initially passed through screens (rows of iron bars, with a spacing of 75–100 cm), which remove large debris, such as wood, paper and bottles. The screens are operated automatically and the screenings are either burnt or macerated and returned upstream of the screens. Grit and small stones are removed either by passing the sewage along a constant velocity channel or through a grit chamber,

Fig. 2.1. An aerial photograph of the Crossness sewage treatment works serving 1.7 million people in 240 km^2 of London (photograph courtesy of Thames Water).

Process flow diagram

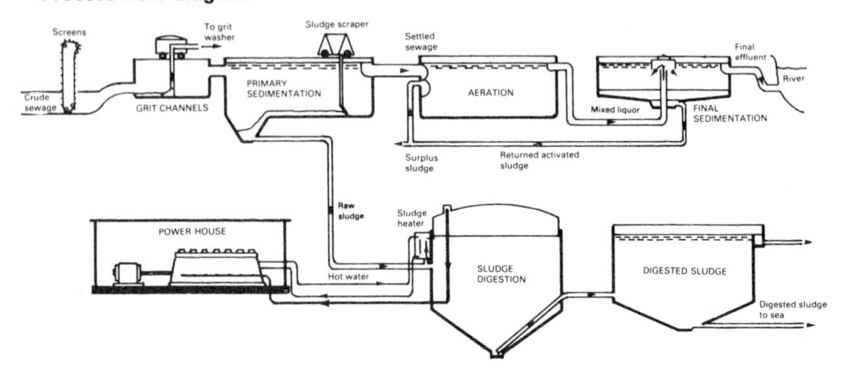

Plan of works

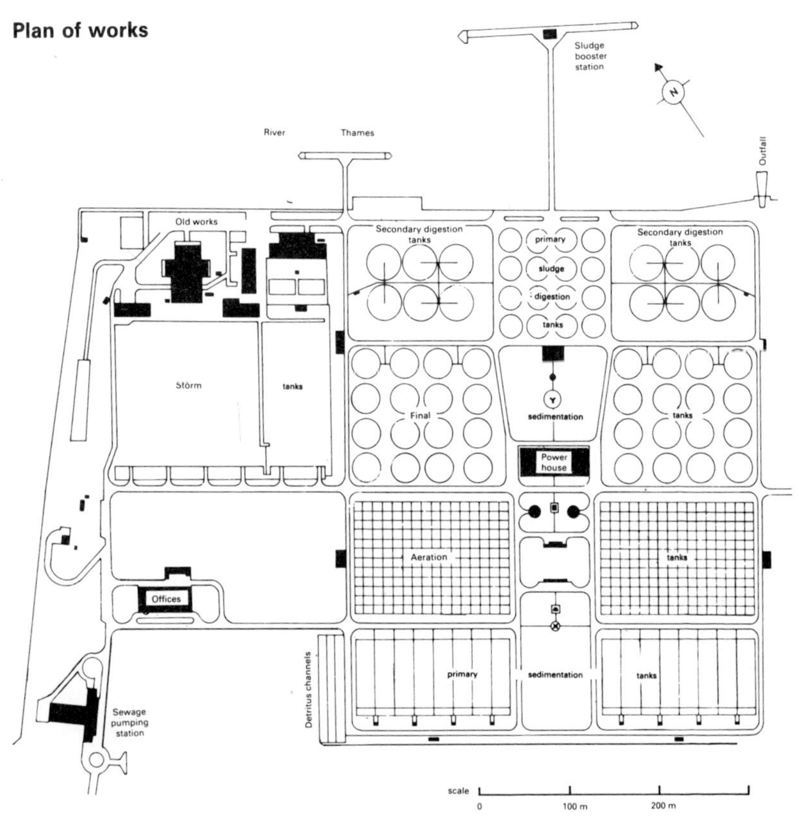

Fig. 2.2. Process flow diagram and plan of works of Crossness sewage treatment works (courtesy of Thames Water).

in which air is introduced from the bottom to create a spiral flow and induce the grit to collect in hoppers. Grit removal may occur before or after screening.

The sewage then passes to the primary treatment tanks.

Primary treatment

The primary treatment or sedimentation process settles out the suspended solids as sludge, which can then be disposed of. There are a variety of designs of sedimentation tank (Austin, 1979) but the ones most frequently installed are of a shallow, radial design, which are equipped with mechanical gear to remove the sludge. The sewage is retained for several hours and about 50 per cent of the suspended solids settle out as primary sludge. Scdimentation is cheaper than biological treatment in terms of unit removal of pollution (Price, 1970) so that the tanks need to be operated at their maximum efficiency.

Secondary (biological) treatment

Secondary treatment involves the oxidation of dissolved and colloidal organic compounds in the presence of micro-organisms and other decomposer organisms. The aerated conditions are obtained usually either by *trickling filters* or *activated sludge tanks*, while in warmer climates *oxidation ponds* may be used. The secondary sludge which results from biological treatment is combined with the primary sludge in sludge digestion tanks, where anaerobic breakdown by micro-organisms occurs.

There are advantages and disadvantages in the methods of biological treatment. Filters tend to be installed in small sewage works, serving populations of less than 50 000. They tend to be higher in capital cost, but lower in running costs, than activated sludge plants (Price, 1970). Filters take up a greater amount of land than activated sludge plants, but they require less skill and active control in functioning. Trickling filters tend to breed flies, which may cause a nuisance to local residents, but activated sludge plants are noisy. Filter beds tend to oxidize more nitrogen than activated sludge plants, but the final effluent carries more suspended solids. Oxidation ponds are much cheaper than other methods of treatment to construct and maintain and they produce a good quality effluent. However, they only function in warm, sunny climates and require large areas of land. The effluent is turbid, due to large populations of algae and the ponds can be the breeding ground of noxious insects, especially mosquitoes.

Trickling (percolating) filters

A cross-section of a typical trickling filter is illustrated in Fig. 2.3. Trickling filters are circular or rectangular tanks, some 1–3 m high and filled with a packed bed of mineral or plastics. The mineral may be broken rock, gravel, clinker or slag but it must be uniformly graded so as to give a large proportion of spaces (voidage). The size range is usually 3.8–5.0 cm, with a specific surface area of 80–110 m^2 m^{-3} of volume and a proportion of spaces of 45–55 per cent of the total volume (Jenkins, 1970; Eden, 1979). Plastics are nowadays widely used instead of minerals because a high voidage can be obtained. The sewage from the primary settling tank is applied to the bed from above, either from rotating arms (circular tanks) or from pipes which travel backwards and forwards over the bed (rectangular tanks). The effluent flowing from the base of the bed contains suspended matter (humus) which is settled out and may be added to the primary settling tank. The clarified effluent may be recycled through the filter so as to dilute the incoming waste water. The main factors influencing the rate of removal of BOD are the specific surface area of the filtration medium, the hydraulic loading and the temperature of the sewage (Pike, 1978).

With the application of settled sewage to the surface of the filter a biological community is gradually established as a slime on the surface of the material and these oxidize the pollutants in the waste

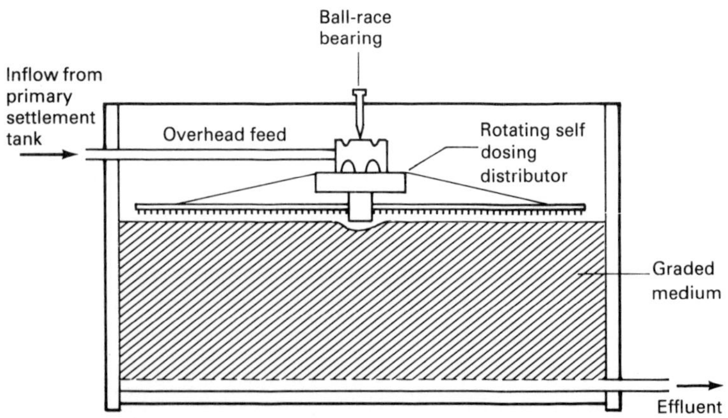

Fig. 2.3. Section through a percolating filter.

water. The ecology of trickling filters has been reviewed by Hawkes (1963). Bacteria are the most numerous organisms and form the base of the food-web. A very wide range of bacteria have been recorded but the dominant aerobic genera appear to be the Gram-negative rods *Zoogloea*, *Pseudomonas*, *Achromobacter*, *Alcaligenes* and *Flavobacterium* (Pike, 1975). Fungi are normally outnumbered 8:1 by bacteria and are most abundant in the top 15 cm (Tomlinson and Williams, 1975). The dominant genera are often *Sepedonium*, *Subbaromyces*, *Ascoidea*, *Fusarium*, *Geotrichium* and *Trichosporon*. These heterotrophic bacteria and fungi are responsible for the primary oxidation of the effluent. Autotrophic bacteria tend to be more predominant in the lower layers of the filter, with *Nitrosomonas* oxidizing ammonium to nitrite and *Nitrobacter* oxidizing nitrite to nitrate.

Algae (e.g. *Chlorella*, *Oscillatoria*, *Ulothrix*) are commonly found in percolating filters but they seem only to play a minor role in the purification process, while in large numbers they may reduce the efficiency of the filter (Benson-Evans and Williams, 1975).

Protozoa are present in proportions similar to fungi and 218 species have been identified, of which 116 were ciliates (Curds, 1975). Ciliates are also usually the most numerous in terms of numbers, including species of *Carchesium*, *Chilodonella*, and *Colpoda*. The major role of protozoa is to remove bacteria which are freely suspended in the liquid film so that the effluent is clarified. Curds *et al* (1968) have examined this experimentally by constructing bench-scale activated sludge plants and the role of protozoa is likely to be the same in percolating filters. In experimental conditions free of protozoa a highly turbid effluent was produced and the turbidity was directly related to the large populations of bacteria suspended in the effluent. BODs were very high. When cultures of protozoa were added the quality of the effluent was improved dramatically as the populations became established.

A diverse grazing fauna is present in percolating filters. It consists of rotifers, nematodes, enchytraeid and lumbricid worms (e.g. *Eiseniella*, *Dendrobaena*), larval and adult Diptera (*Anisopus*, *Psychoda*, *Metriocnemus*) as well as Coleoptera and Collembola. Williams and Taylor (1968) have shown that filters with macro-invertebrates produce a much better effluent than those without. This is partly due to the grazing activity of the animals, which prevent the accumulation of too much film and increase the oxygen diffusion. The film material is made more settleable by passing it through the animals' guts. Material which has been converted to chitin of exuviae of larval and pupal *Psychoda* settles quickly and a considerable proportion of the humus solids in percolating filters may consist of chitin (Solbé *et al*, 1967). Finally Solbé and Tozer (1971) and Solbé (1971) have shown that the respiration of the macro-invertebrate community may be responsible for between 3 per cent (in winter) and 10 per cent (in summer) of the total carbon dioxide dissipated and this is a valuable contribution to the purification of sewage in biological filters.

To function efficiently, the percolating filter requires a continuous inoculation of micro-organisms in the sewage. To assist in the establishment of an active film, a new filter bed may be seeded with humus sludge or activated sludge solids. Establishment is most rapid in the summer, but it may be up to two years before the community is fully developed.

The community developing will depend on the depth within the filter bed, the time of the year and the composition of the waste to be treated (Pike, 1975). Fungi, for instance, do better than bacteria at low temperatures, in acid media and in waste waters with a high organic content (Pike, 1978). The rate of accumulation of the film is determined by the difference between the rate of growth and rate of removal and its control is important in the efficient functioning of a filter bed. A thick film impedes the flow of water through the bed and the growth rate of bacteria may be reduced because the rate of diffusion of nutrients through the slime decreases. Recirculation of effluent after it has passed through the filter dilutes the incoming effluent, thus reducing the strength of feed, and increases the hydraulic load and these both reduce the rate of growth of the film. Alternate double filtration is frequently used, in which two filters are operated in series. The waste water is applied at a relatively high rate to the primary filter and its effluent, after settlement, is passed to the second filter. The filters are switched at daily or weekly intervals. The primary filter quickly grows a thick film, but when it is switched to the secondary position the film rapidly shrinks. The overall costs are generally lower than for a single filtration. The biology of film control has been discussed in detail by Hawkes (1963).

The kinetics of microbial growth and the oxidation of substrates in the percolating filter can be approximately described by the Monod equation. The rate of increase of biomass per unit of biomass is known as the specific growth rate (μ) and this can be related to the growth limiting concentration (S) of the substrate:

$$\mu = \frac{\mu m\ S}{Ks + S} \qquad [2.1]$$

where μm is the maximum specific growth rate for that limiting substrate and Ks is the saturation constant (the concentration of the limiting substrate at which $\mu = \frac{1}{2}\mu m$).

If the yield (Y), the mass of cells produced in the consumption of a unit biomass of substrate, is constant, then the growth rate can be related to the substrate removal:

$$\frac{dS}{dt} = \frac{1}{Y}\frac{dx}{dt} = \frac{1}{Y}\frac{\mu m\, Sx}{Ks + S} \qquad [2.2]$$

These equations have been widely used in describing the rate of removal of complex substances (e.g. BOD) in the sewage treatment process, but it must be remembered that the Monod equation assumes that the growth of a single strain of a micro-organism is controlled by a single substrate, all others being in excess. In the treatment process many species of micro-organisms are simultaneously decomposing many substances. However, the first-order kinetics are approximated and the Monod equation provides a basis from which more complex models can be built.

Activated sludge process

Activated sludge tanks can be seen in Fig. 2.1. The settled sewage is mixed with a flocculent suspension of micro-organisms and aerated in a tank for from 1 to 30 hours, depending on the treatment required. The medium is rich in dissolved and suspended nutrients, rich in oxygen and violently agitated. The organic matter is removed by oxidation, adsorption and flocculation, and sludge, which increases by 5–10 per cent during the process, is removed from the purified effluent in a sedimentation tank and then returned to the inlet of the aeration tank. Any excess is returned to the inlet of the primary sedimentation tank.

Hawkes (1963) has pointed out that, by contrast to the percolating filter, the activated sludge is a truly aquatic environment. The turbulent conditions within the tank are unsuitable for macro-invertebrates so that the community lacks the higher links in the food web. The amount of microbial mass in the system is controlled by withdrawing excess sludge, whereas in the filter excess film is removed chiefly by biological agencies. In the activated sludge tank the microbial community is initially associated with the untreated waste and finally with the purified effluent, whereas in the filter bed a succession of communities are established at different depths in the bed and are associated with different degrees of effluent purification (Hawkes, 1963).

The majority of bacteria described in the activated sludge belong to Gram-negative genera (Pike, 1975), such as *Pseudomonas*, *Zoogloea* and *Sphaerotilus*. Fungi are not usually dominant in the activated sludge process, though they may grow profusely if bacteria are inhibited, often when industrial effluent is present in the sewage. Fungi may cause 'bulking' in the sludge, in which loose-flocculent growths of micro-organisms impede settling.

Curds (1975) has listed 228 species of protozoans associated with activated sludge, of which 160 were ciliates. Activated sludge thus has a greater species richness than percolating filters, but a much greater proportion of them are ciliates. Densities of protozoa are of the order of 50 000 cells ml^{-1}. Hawkes (1963) has described a generalized succession of protozoans. Rhizopods and flagellates, able to utilize dissolved and particulate wastes, predominate initially, but later, as bacteria increase, predatory flagellates and free-swimming ciliates are dominant. Crawling and attached ciliates predominate in the later stages of floc formation. The role of protozoa is, as in the filter, to clarify the effluent.

Small numbers of nematodes and rotifers are also present in activated sludge.

For efficient operation, the activated sludge process requires that the concentrations of substrate and micro-organisms should be low. The detailed operation of activated sludge and percolating filter plants is described by Klein (1966).

Sludge digestion and disposal

The sludge produced by primary and secondary treatment processes is passed to sludge digestion tanks, where it is decomposed anaerobically. Alternatively, it may be stored for later disposal. The sludge amounts to some 50 per cent of the initial organic matter entering the sewage works. The role of anaerobic bacteria is to convert the highly putrescible raw sludge into a stable and disposable product, which neither gives rise to offensive smells nor attracts harmful insects or rodents (Crowther and Harkness, 1975). The major constituents of raw sludge are proteins, fats and polysaccharides and they are degraded in three processes. Hydrolysis involves the formation of long chain fatty acids, amino acids, monosaccharides and disaccharides. The second process, acid formation, results in the production of a range of fatty acids, alcohols, aldehydes and ketones, together with ammonia, carbon dioxide, hydrogen and water. The third process, methanogenesis, results in methane, carbon dioxide and water. There is also, of course, an increase in bacterial biomass. The biochemistry of methanogenesis is outlined by Crowther and

Harkness (1975) and Zehnder (1978).

Sludge digestion often takes place in two stages (Klein, 1966). In the first stage, digestion takes place in closed tanks, heated to a temperature of 27–35 °C for 7–30 days and most of the gas evolution occurs here. Some of the gas evolved is used to heat the tanks. In the second stage, further digestion occurs for 20–60 days in open tanks at ordinary temperatures and the sludge consolidates and dries with the separation of a supernatant liquor, which is then returned to the sewage inlet for treatment with the sewage.

The sludge, after digestion, has been reduced in volume by two thirds and its disposal presents some problems. Sludge may be dumped at sea, it may be incinerated or it may be used as a land-fill. It is also a potentially valuable fertilizer and some 40 per cent of sludge in England and Wales is returned to the land. The sludge, however, may contain toxic materials, such as metals or fluoride and these could affect productivity. Wollan *et al* (1978) found that germination of barley and rye grass was inhibited in the presence of sewage sludge. They established that this was due to the production of inhibitors such as ethylene and ammonia, caused by the lowering of oxygen tensions in the soil following intense microbial activity. However, sewage sludge is a good fertilizer (e.g. Coker, 1966) and improves soil aggregation (Epstein, 1975) and, with the soaring cost of fertilizers, it is economically attractive to farmers, particularly in situations where transport costs are low. Prospects and problems are reviewed in Bruce (1979).

There has been considerable interest recently in methods to recover or produce compounds from organic wastes and these have been summarized by the Water Research Centre (1974). They involve direct utilization of wastes, for example for animal feeds, the direct extraction of proteins, fats and other organic compounds and the microbial conversion of carbohydrate to protein. The development of genetic engineering techniques may enable many valuable materials to be recovered or synthesized from domestic or industrial wastes.

Oxidation ponds

Oxidation ponds are used in warm climates to purify sewage and the process involves an interaction between bacteria and algae (Fig. 2.4). The ponds are shallow lagoons, with an average depth of 1 m. Settled sewage passes through the pond in 2–3 weeks, but raw sewage may be retained for up to 6 months. The bacteria in the pond decompose the biodegradable organic matter to release carbon

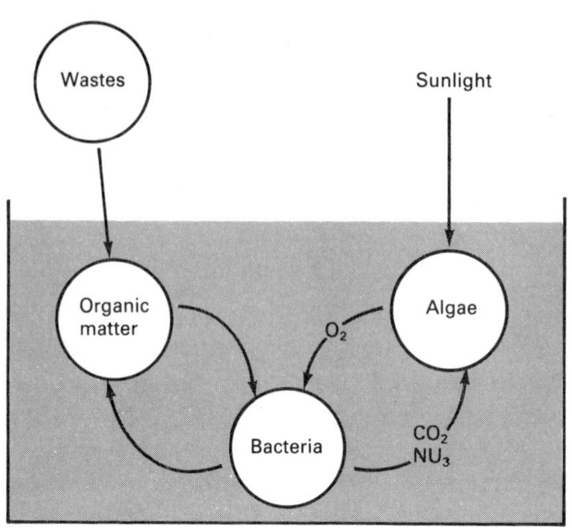

Fig. 2.4. Summary of processes in an oxidation pond.

dioxide, ammonia and nitrates. These are utilized by the algae, together with sunlight, and the photosynthetic process releases oxygen, enabling the bacteria to break down more waste.

Oxidation ponds not only purify wastes, but they can also be used to provide energy. The algae have a solar conversion efficiency of 3–5 per cent and when harvested they may be fed directly to animals, fermented to produce methane or burnt to produce electricity (Hall, 1979). Fermentation to produce methane can yield 11.6 MJ kg^{-1} algae of energy and this could produce 5 per cent of local methane usage in the USA, or 10 per cent if animal wastes are included. Fertilizers (N and P) are produced as a by-product and one unit area of oxidation pond could fertilize 10–50 areas of agriculture (Hall, 1979).

Yields of 50–60 tonnes (dry) ha^{-1} y^{-1} of algae, a biomass easily obtained in Californian oxidation ponds, would produce 74 000 kWh of electricity. Oswald (1977) has described how bio-conversion of wastes in oxidation ponds adjacent to cattle and chicken farms can produce a protein rich food which is re-fed to the animals. About 40 per cent of the nitrogen in the waste is recovered in the algae. Oswald has calculated that 4 m. hectares of oxidation ponds could provide for the protein requirements of the United States, compared with 121 m. hectares of traditional agricultural land. If fish were harvested from the oxidation ponds the protein conversion efficiency would no doubt be considerably greater.

Tertiary treatment

In many situations the dilution available to an effluent in the receiving water is insufficient to prevent a deterioration of water quality. To prevent this a higher quality effluent must be produced and this effluent 'polishing' is known as tertiary treatment. The removal of nitrates and phosphates is described in Chapter 3. The most common method of polishing effluents is by passing them through rapid sand filters, reviewed by Jago (1977), or by micro straining, resulting in a final effluent with a BOD of less than 10 mg l^{-1} and a similar level of suspended solids. Secondary effluent may also be passed over grassland or through marshes where these are available, (e.g. Sloey *et al*, 1978). Where sufficient land is available the secondary effluent may be run into shallow lagoons to mature, resulting in a high quality effluent (Potten, 1972). These lagoons are very eutrophic and the growth of fish in them is extremely rapid (White and Williams, 1978) and by harvesting commercially valuable fish from stabilization lagoons, considerable savings on the cost of the overall treatment process could be obtained (Wert and Henderson, 1978).

Removal of pathogens

The reduction in the numbers of pathogens during the sewage treatment process is governed by the length of retention time during treatment, chemical composition of the wastes and their state of degradation, antagonistic forces in the biological flora, pH and operational temperature, together with other less understood factors (Geldreich, 1972). Trickling filters have been found to reduce densities of *Salmonella paratyphi* by 84–99 per cent, enteric virus by 40–60 per cent and cysts of *Entamoeba histolytica* by 88–99 per cent while waste stabilization lagoons result in further losses of pathogens. With further considerable dilution in the receiving water the sewage treatment process can be seen as an efficient way of reducing the incidence of pathogens. Greater potential problems are associated with home septic tanks, which are often overloaded and inefficiently maintained and also with storm drainage.

Conclusions

This section has largely been concerned with the treatment of sewage, but the principles apply equally well to the treatment of industrial organic wastes. Many factories have small-scale effluent treatment plants to deal with their organic wastes before these are discharged into rivers.

Sewage treatment has been described in detail because it demonstrates how biological systems can be applied effectively to industrial processes. A study of these processes can in turn help in the development of new fundamental biological theory, for instance in understanding the dynamics of ecological communities. The discussion has also shown that there are many potential by-products of economic importance to be gained from the treatment process, such as energy, organic compounds or food, either directly as fish or indirectly as animal feed, and the exploitation of this potential will involve the skilled manipulation of biological systems.

In an ideal world, all of the materials in sewage would be economically recovered and water alone would be returned into the hydrological cycle. The world, of course, is not ideal. Even where sewage is receiving secondary treatment, the effluent produced may be substandard because the design capacity of the works may be below the population it is now having to serve and resources are not available to expand the plant. Similarly, many plants are old and inefficient or they may become overloaded during unusual conditions, e.g. storms. Sewage is also changing with time (e.g. with increased amounts of oils and plastics) and these may present problems with which old plants are not designed to cope. In addition to sewage, other forms of organic pollution may be affecting the receiving water and many parts of the world have no effective waste treatment facilities at all. The next section will outline the effects of organic pollution on the receiving stream.

Effects of organic pollutants on receiving waters

The oxygen sag curve

When an organic polluting load is discharged into a river it is gradually eliminated due to the activities of micro-organisms, by methods very similar to those occurring in the sewage treatment process. This *self-purification* requires sufficient concentrations of oxygen and complex organic molecules are broken down into simple inorganic molecules. Dilution, sedimentation and sunlight also play a part in the process. Esser (1978) has shown that attached micro-organisms in streams play a greater role than suspended organisms in self-purification and this role increases as the quality of the effluent increases because attached micro-organisms are already present in the stream, whereas suspended micro-organisms are mainly supplied with the discharge.

The de-oxygenation of a river caused by organic wastes is generally a slow process so that the point of maximum de-oxygenation may occur considerably downstream of a discharge. The degree of de-oxygenation depends on a number of factors (Klein, 1962), such as the dilution that occurs when the effluent mixes with the stream, the BOD of the discharge and of the receiving water, the nature of the organic material, the total organic load in the river, temperature, the extent to which re-aeration occurs from the atmosphere, the dissolved oxygen in the stream and the numbers and types of bacteria in the effluent.

Figure 2.5 illustrates generalized curves, known as *oxygen sag curves*, which are obtained when dissolved oxygen is plotted against the time of flow downstream. The curves show slight, moderate and gross pollution. A poor quality effluent contains a high concentration of ammonia and the eventual oxidation of this to nitrite and nitrate may cause a further sag in the oxygen curve.

Figure 2.6 illustrates a real example of an oxygen sag curve, that measured during the four quarters of the year 1967 in the River Thames below Teddington Weir (Arthur, 1972). Note that the shape of the sag curve (the downstream position of the minimum oxygen and the length of river over which low oxygen conditions prevail) is largely determined by the amount of water flowing over Teddington Weir. With low flows in the autumn, minimum oxygen conditions prevail over a distance of some 40 km downstream from London Bridge, whereas with high flows in the first quarter of the year only 12 km have minimum oxygen and this situation is not reached until 22 km below London Bridge.

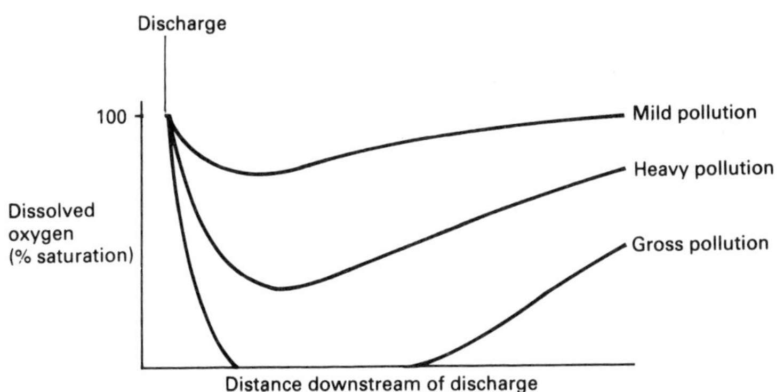

Fig. 2.5. The effect of an organic discharge on the oxygen content of river water.

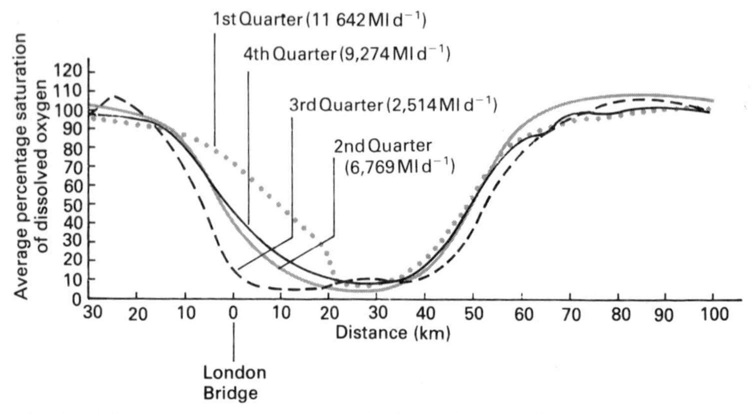

Fig. 2.6. The oxygen sag curve in the tidal stretch of the River Thames in the four quarters of 1967–average daily flow over Teddington weir in brackets (adapted from Arthur, 1972).

If the oxygen demand of a sewage effluent is studied over a number of days, it is found that oxidation proceeds quite rapidly to begin with, but then slows down over a period of 15–20 days. However, there are often two further stages in oxidation, which may account for a large proportion of the total oxygen consumption. The oxygen demand in the first 20 days is due to the oxidation of organic matter (carbonaceous BOD), while later demand involves the oxidation of ammonia to nitrite and then nitrate.

The rate of biochemical oxidation of organic matter by micro-organisms (k_1) was considered by Streeter and Phelps (1925) to be proportional to the remaining concentration of unoxidized substance. This reaction follows a typical monomolecular curve and can be represented:

$$-\frac{dL}{dt} = K_1 L \qquad [2.3]$$

where L is the ultimate carbonaceous demand at any time, t, and K_1 is the rate constant of oxidation (deoxygenation constant).

Integrated, the equation becomes:

$$L = L_0 e^{-K_1 t} \qquad [2.4]$$

where L_0 is the initial value of L.

This can be written in a more convenient form as:

$$y = L_0(1 - e^{-K_1 t}) \qquad [2.5]$$

where y is the uptake of oxygen in time $t(= L_0 - L)$ or, converting to base 10 logs:

$$y = L_0(1 - 10^{-k_1 t}) \qquad [2.6]$$

For sewage effluents, k_1 is about 0.1 at 20 °C. The daily uptake of oxygen ($t = 1$) is therefore:

$$y = 0.206\ L_0 \qquad [2.7]$$

or about 20 per cent of the oxidizable matter present in the sewage will be oxidized the following day.

The rate of the BOD reaction is a function of temperature and Streeter and Phelps (1925) expressed the effect of temperature by the equation:

$$(k_1)_T = (k_1)_{20}\theta^{(T-20)} \qquad [2.8]$$

where $(k_1)_T$ is the value of the deoxygenation constant at any temperature T °C, $(k_1)_{20}$ is the value at 20 °C and θ is a temperature coefficient.

A number of curve fitting techniques are available to determine the reaction rate k_1 and these are described in detail in Nemerow (1974).

The first order kinetics of the equations described above may be somewhat complicated by the occurrence of nitrification simultaneously with carbonaceous oxidation. There may also be a lag phase in the early part of the BOD curve, which can be overcome by substituting $(t - t_0)$ in place of t, where t_0 is the duration of the lag phase, in the above equations.

This monomolecular approach to understanding the biochemical oxidation of sewage tends to give a poor fit to the observed results, particularly in the later stages of the process. Many factors will affect the performance of bacteria on this complex substrate. The chemical form of the substrate and whether it is in solution or suspension will be important (Klein, 1962). Respiration of bacteria, and hence their oxygen requirements, will be higher during growth than during the resting stages. To overcome some of these difficulties, Gameson and Wheatland (1958) have utilized a retarded exponential, which assumes that all the constituents of sewage are oxidized at the same rate, but that the rate constant decreases with time:

$$y = L_0[1 - (1 + at)^{-K_1/a}] \qquad [2.9]$$

where a is the coefficient of retardation and K_1 is a measure of the initial rate of oxidation.

Gameson and Wheatland (1958) have also produced a composite equation which recognizes that the different components of sewage will oxidize at different rates:

$$y = L[1 - (p_1 e^{-Kp_1 t} + p_2 e^{-Kp_2 t} + \cdots p_n e^{-Kp_n t})] \qquad [2.10]$$

where p_1, p_2, p_n are the proportions of the various components present, which have rate constants Kp_1, Kp_2, Kp_n.

The oxygen demand by organic effluents has been described above, but oxygen is also replaced in the stream. If pollutional conditions are slight enough to allow the growth of green plants some re-oxygenation will occur by photosynthesis. Re-aeration will also occur by diffusion from the atmosphere and the rate of aeration is proportional to the oxygen saturation deficit.

Streeter and Phelps (1925) developed the following equation:

$$\frac{dD}{dt} = K_1 L - K_2 D \qquad [2.11]$$

where D is the oxygen deficit and L the ultimate carbonaceous demand. The deoxygenation reaction is represented by ($K_1 L$) and the re-aeration reaction by ($K_2 D$).

Integrating:

$$D = \frac{k_1 L_0}{k_2 - k_1}[10^{-k_1 t} - 10^{-k_2 t}] + D_0 \cdot 10^{-k_2 t} \qquad [2.12]$$

where L_0 and D_0 are respectively the initial BOD and oxygen deficit in the stream and k_1 and k_2 are the rate constants for de-oxygenation and re-aeration. This is the *sag equation*.

A number of variables affect the rate of re-aeration, such as velocity, depth, slope and channel irregularity. Re-aeration also increases with temperature. There are a number of ways of determining the re-aeration rate (k_2) and these are discussed in Nemerow (1974). It is possible from these equations to determine the oxygen deficit at any site below a discharge.

To increase the rate of aeration and speed up self-purification below discharges, weirs are often built into rivers to considerable effect. Recently, to combat the disastrous effects of accidental pollution incidents, techniques have been developed to pump pure oxygen directly into threatened rivers (Anon., 1979). It is essential that the pollution be spotted in time, but the technique can be especially valuable where prime fisheries are at risk.

Effects on the biota

Organic pollution affects the organisms living in a stream by lowering the available oxygen in the water, and hence causing reduced fitness, or, when severe, asphyxiation, by increasing the turbidity of the water and hence reducing the light available to photosynthetic organisms, and by settling out on the bottom of the stream, thus altering the characteristics of the substratum. The general effects of fairly severe organic pollution are illustrated in Fig. 2.7 following Hynes (1960). The top part of the figure (A) shows the oxygen sag curve, together with a massive increase in BOD, salts and suspended solids at the point of discharge, followed by a gradual decline in these parameters as self-purification occurs. The peak of ammonia (B) is replaced by a peak of nitrate as nitrification proceeds and both are gradually diluted as the polluting load travels downstream.

There is a large increase in the number of bacteria (C) immediately below the outfall and these gradually decline as their substrate is depleted. There is also a large increase in sewage fungus (see later), which disappears as the stream re-oxygenates. The protozoa are chiefly predators on bacteria and increase in response to bacterial increases, subsequently to decline as bacterial numbers fall. Algae, and especially *Cladophora* (blanket weed, a large filamentous green alga) increase in numbers as recovery begins, light conditions improve and nutrients are released from the oxidizing organic matter. They decrease as nutrients are used up or diluted.

The stream animals are shown at the bottom of Fig. 2.7. The clean water fauna is eliminated at the point of discharge of the pollutants, unable to tolerate the lowered oxygen tension. Sludge worms (Tubificidae) may be the only macro-invertebrates present immediately below the discharge and, in some cases of very severe pollution, even these may be absent. As conditions improve blood worms, larvae of the midge *Chironomus*, become abundant and, as further amelioration occurs, large populations of the isopod *Asellus* build up. As the stream gradually re-oxygenates, the clean water fauna increases in numbers and diversifies.

Micro-organisms

The primary effect of organic pollution is to increase the numbers of bacteria which use the waste as a substrate. In the River Danube, for instance, Deufel (1972) showed that bacteria increased markedly below the sewage outfalls of large cities, with direct counts of up to 36×10^6 ml^{-1}. Edwards and Owens (1965) considered that most of

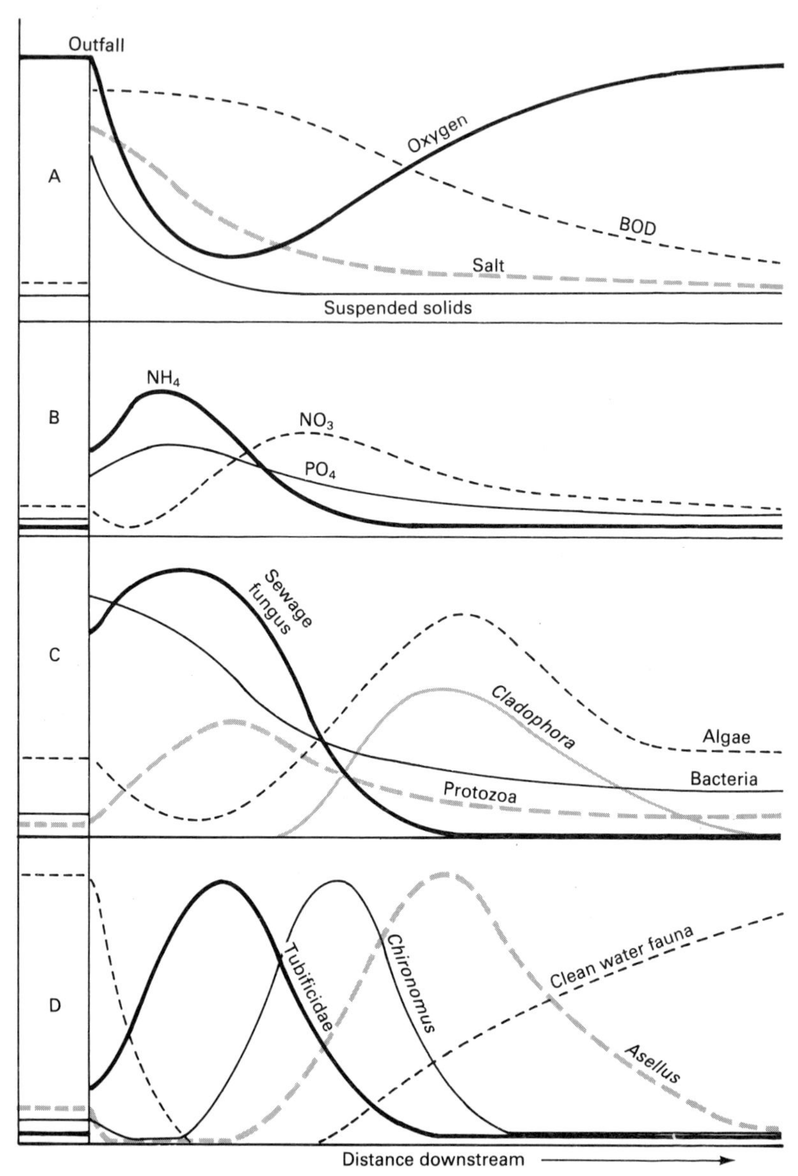

Fig. 2.7. Schematic representation of the changes in water quality and the populations of organisms in a river below a discharge of an organic effluent (from Hynes, 1960).

A Physical changes
B Chemical changes
C Changes in micro-organisms
D Changes in macro-invertebrates

the bacteria below an outfall were suspended, with 10 per cent being attached to plant surfaces.

Under fairly heavily polluted conditions a benthic community of micro-organisms, known as *sewage fungus*, develops. Sewage fungus is an attached macroscopic growth, which may form a white or light brown slime over the surface of the substratum, or may exist as a fluffy, fungoid growth, with long streamers (Hawkes, 1962). It is not fungi, but usually bacteria, which dominate the sewage fungus community. Butcher (1932) carried out some early studies on the community and Curtis and Curds (1971) have more recently examined in detail the organisms making up sewage fungus, a list of which is given in Table 2.2. Some of these organisms are illustrated in Fig. 2.8. The dominant species are usually *Sphaerotilus natans* and zoogloeal bacteria.

Sphaerotilus natans consists of an unbranched filament of cells, enclosed in a sheath of mucilage. The sheath enables the bacterium to attach to solid surfaces, as well as protecting the organism from parasites and predators (Venosa, 1975). These bacteria use a wide variety of organic compounds as substrates and they can also utilize inorganic nitrogen sources, though growth is less luxuriant than with

Table 2.2. The typical organisms of the sewage fungus community (from Curtis and Curds, 1971).

Bacteria	*Sphaerotilus natans*
	Zoogloeal bacteria
	Beggiatoa alba
	Flavobacterium sp.
Fungi	*Geotrichum candidum*
	Leptomitus lacteus
Protozoa	*Colpidium colpoda*
	Colpidium campylum
	Chilodonella cucullulus
	Chilodonella uncinata
	Cinetochilum margaritaceum
	Trachellophyllum pusillum
	Paramecium caudatum
	Paramecium trichium
	Uronema nigricans
	Hemiophrys fusidens
	Glaucoma scintillans
	Carchesium polypinum
Alga	*Stigeoclonium tenue*
	Navicula spp.
	Fragilaria spp.
	Synedra spp.

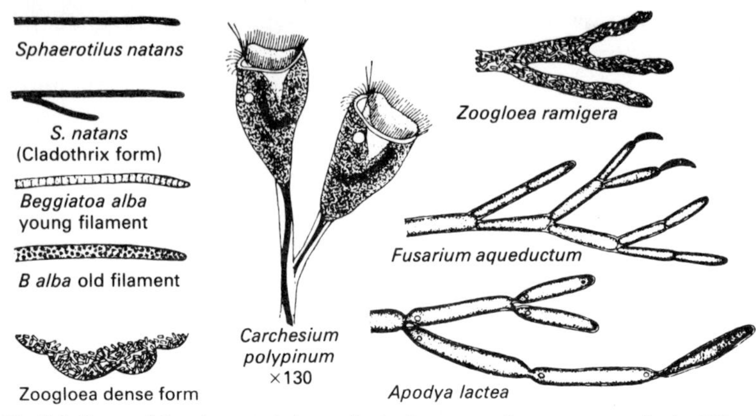

Fig. 2.8. Some of the characteristic species in the sewage fungus community (×330, except *Carchesium*) (from Hynes, 1960).

organic nitrogen sources (Stokes, 1954; Mulder and Veen, 1963; Veen *et al*, 1978). Curtis *et al* (1971) found that, using glucose and acetate as carbon sources in artificial channels, slime did not form below a concentration of 1 mg l^{-1}, while above this concentration slime formation was proportional to oxygen concentration. The bacteria are aerobes, but they can survive in oxygen concentrations down to 2 mg l^{-1}. For typical growth, a certain amount of flow is also necessary. The sewage fungus community extends further downstream during the winter than in summer, because the oxidation of organic matter in the effluent proceeds more slowly, such that the pollution plug extends further downstream, and also because the sewage fungus bacteria are able to compete more effectively with other heterotrophic bacteria at lower temperatures.

Eleven protozoan species occur in very large numbers in the sewage fungus community (Curtis and Curds, 1971, Table 2.2). *Carchesium*, a stalked, ciliated protozoan, tends to be most abundant at the lower end of the sewage fungus zone, where oxygen conditions are somewhat improved. It is a predator, feeding mainly on the large populations of bacteria present in sewage effluent. *Carchesium*, when present in large numbers, can cause silting of the river bed because it flocculates suspended matter (Sugden and Lloyd, 1950). *Colpidium colpoda* and *Chilodonella cucullulus*, both holotrichs, were recorded by Curtis and Curds (1971) to be very abundant in the sewage fungus. *Colpidium* feeds mainly on bacteria, while *Chilodonella* is a more generalized feeder on bacteria, diatoms, filamentous growths and flagellates.

Algae

With heavy organic pollution, de-oxygenation and low light, algae are eliminated from rivers, but there is a gradual re-appearance as conditions improve and populations and growth are stimulated by the large concentrations of nutrients present. Butcher (1947) examined the algae of organically enriched rivers by looking at species growing on immersed glass slides and the frequency of species in two rivers is shown in Fig. 2.9. The filamentous *Stigeoclonium tenue* became common in the zone immediately below the region of gross pollution. *Stigeoclonium* also occurs in the sewage fungus community (Table 2.2). The characteristic species of the recovery zone were the diatoms *Nitzschia palea* and *Gomphonema parvulum.* The blue-green *Chamaesiphon* sp., the green *Ulvella frequens* and the diatom *Cocconeis placentula* appeared when the pollution had dispersed. A long distance is required for the complete recovery of the algal community in grossly polluted rivers; this amounted to 70 km downstream of the discharge on the River Tame and 56 km on the River Trent.

The diversity of algal species in clean waters can be very variable, according to Archibald (1972), who studied South African streams.

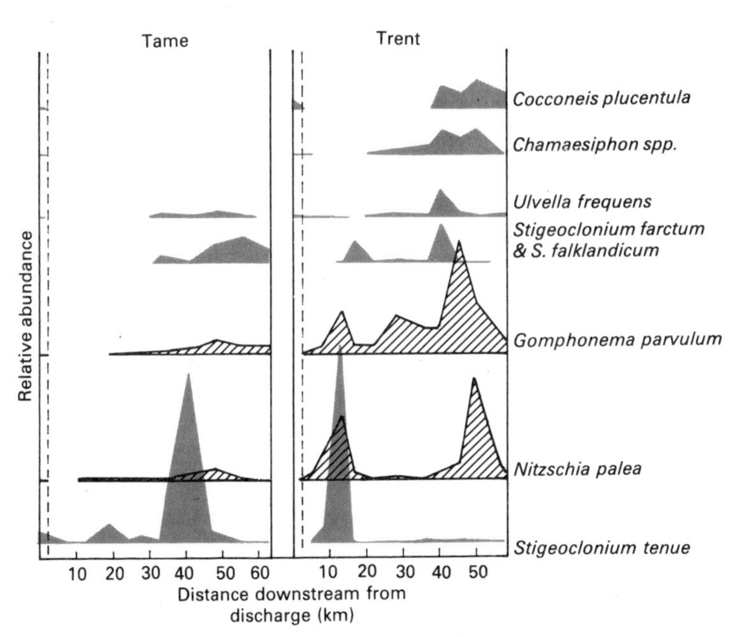

Fig. 2.9. The principal composition of the algal flora growing on glass slides placed at sites downstream of an organic discharge in two English Midland rivers (adapted from Hynes, 1960, from data in Butcher, 1947).

Heavily polluted environments always have communities low in numbers of species because sensitive species are gradually eliminated as the pollution load increases. However, at low levels of organic pollution, tolerant species tend to increase, whilst conditions are not sufficiently severe to cause a great loss of sensitive species, so that mildly polluted rivers can have high diversity (Fig. 2.10). It is therefore essential, in equating diversity with pollution load, to examine which species are present and have a knowledge of their requirements. For instance, Archibald (1972) recorded that *Achnanthes minutissima* dominated a low-diversity community of diatoms in the Vaal River, but as this species requires highly oxygenated water for good growth, the low community diversity could not have been caused by organic pollution.

The filamentous green alga, *Cladophora*, becomes abundant in the recovery zone and forms dense blankets over the substratum, where it provides both cover and food for invertebrates. The ecology of the plant has been reviewed by Whitton (1970). Its dense growth in the recovery zone is linked to an increase in nutrients, and especially to phosphate present at concentrations greater than 1 mgP l^{-1} (Pitcairn and Hawkes, 1973).

Higher plants

Like algae, higher plants (macrophytes) are adversely affected by the de-oxygenation of water with organic pollution, by the load of suspended solids which reduce light and by the settling out of solids which may render the bed of the river unstable. However, when conditions improve downstream, the rich supply of nutrients can

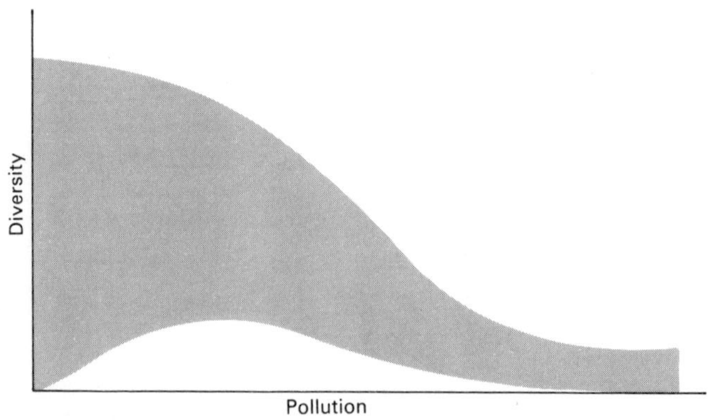

Fig. 2.10. The relationship between the diversity of diatoms and polluting load (from Archibald, 1972).

result in a very dense growth of some species. Haslam (1978) has described the effects of organic pollution on plants in rivers. She records a decrease and loss of those species most sensitive to pollution, with a decrease in overall species richness, and an increase in any species favoured by pollution. The only species whose range appears to be increased by organic pollution is *Potamogeton pectinatus*, described by Haslam (1978) as very tolerant. Five species of higher plant were described as tolerant, being common in both clean and polluted streams, viz. *Mimulus guttatus*, *Potamogeton crispus*, *Schoenoplectus lacustris*, *Sparganium emersum* and *S. erectum*. A further seven species were fairly tolerant.

Figure 2.11 illustrates the number of macrophyte species in the upper reaches of the River Trent catchment in the English Midlands. *Potamogeton pectinatus* occurs alone below the Potteries towns (including Stoke-on-Trent), from which the river receives treated sewage effluents, but the plant grows luxuriantly. The unpolluted Rivers Sow and Penk, relatively rich in macrophytes, dilute the polluted water of the Trent and the number of species increases downstream.

Overlying the effects of pollution on macrophytes are many natural characteristics of rivers, such as substratum type, flow regime, current speed, water chemistry and shading. These factors make it difficult to relate a particular plant community to pollutional conditions, unless neighbouring streams with similar characteristics, but different loads of pollutants, are compared.

Invertebrates

Protozoa dominate the animal community in polluted rivers where oxygen is severely limiting and some species have already been mentioned in the discussion of sewage fungus. The response of protozoan communities to pollution is difficult to evaluate because large numbers are washed into streams from the surrounding land and from the secondary sewage treatment process in the effluents, making it a problem to distinguish between natural and intrusive members of the fauna.

There has been a great deal of work on the effects of organic pollution on macro-invertebrate communities. Both Hynes (1960) and Hawkes (1962) comment that heavy pollution affects whole taxonomic groups of macro-invertebrates, rather than individual species. Specific differences only become important in cases of mild pollution. In general those organisms associated with the silted regions of rivers are the most tolerant of organic pollution, while species associated with eroding substrata and swiftly flowing water

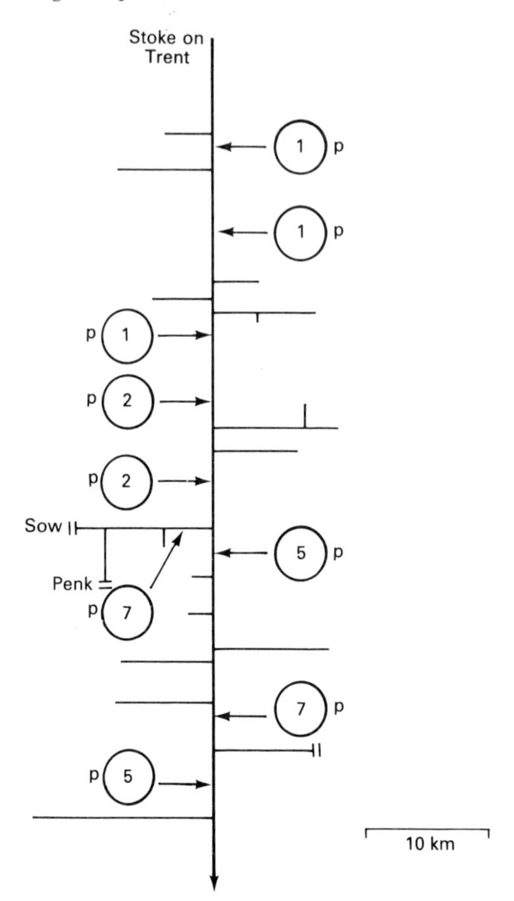

Fig. 2.11. The number of species of aquatic macrophytes in the upper River Trent, receiving heavy pollution from Stoke-on-Trent. p = the pollution tolerant species *Potamogeton pectinatus* (adapted from Haslam, 1978).

are the most sensitive. Invertebrates from swiftly flowing waters generally have higher metabolic rates than those from slow flowing waters and they would hence tend to be more sensitive to decreases in oxygen content in the water. Respiratory adaptations in invertebrates are discussed by Hynes (1970). Some species tolerant and intolerant of organic pollution are illustrated in Fig. 2.12 and a summary of the invertebrate taxa associated with different degrees of organic enrichment has been given by Hawkes (1979).

In heavily polluted waters tubificid worms are frequently very abundant, often forming monocultures with densities of over 10^6 m^{-2} (Brinkhurst, 1970). Successive species of tubificids are eliminated as

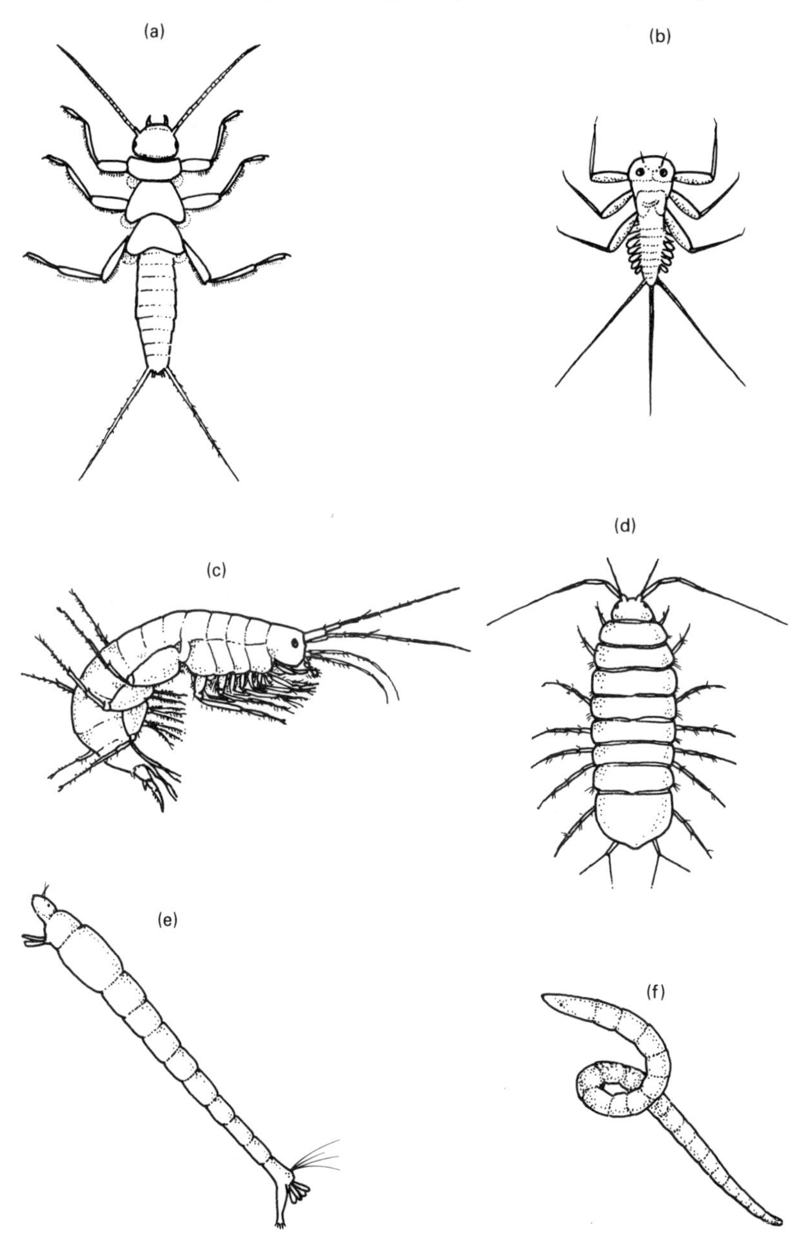

Fig. 2.12. Some species of freshwater invertebrates, in order of increasing tolerance to organic pollution (a) *Dinocras cephalotes* (Plecoptera); (b) *Ecdyonurus venosus* (Ephemeroptera); (c) *Gammarus pulex* (Amphipoda); (d) *Asellus aquaticus* (Isopoda); (e) *Chironomus riparius* (Diptera); (f) *Tubifex tubifex* (Oligochaeta) (adapted from Macan, 1959)

conditions become more severe and in very severe conditions only *Limnodrilus hoffmeisteri* and/or *Tubifex tubifex* remain. For these animals the effluent provides an ideal medium for feeding and burrowing, while the absence of predators allows populations to increase largely unchecked.

The success of tubificids in these environments is due to their ability to respire at very low oxygen tensions. Figure 2.13 shows that the respiration rate of *Tubifex tubifex* and *Branchiura sowerbyi* is almost unaffected by dissolved oxygen concentrations down to 20 per cent of air saturation. Tubificids contain the pigment haemoglobin, which has a high affinity for oxygen. The haemoglobin in *T. tubifex* exhibits a negative Bohr effect, which means that oxygen can be taken up at low pH (Palmer and Chapman, 1970), when the carbon dioxide content of the water is high, a frequent occurrence in organically polluted waters. The pigment functions in the transport of oxygen, but it appears not to store oxygen for use during prolonged periods of anoxia. Tubificids are known to survive in anaerobic conditions for up to four weeks and they may metabolize glycogen anaerobically at this time (Dausend, 1931). Tubificids can also feed, defaecate and lay eggs at low oxygen tensions (Aston, 1973).

Naidid worms may also respond to organic pollution by large increases in numbers, especially on the stony substrata which are

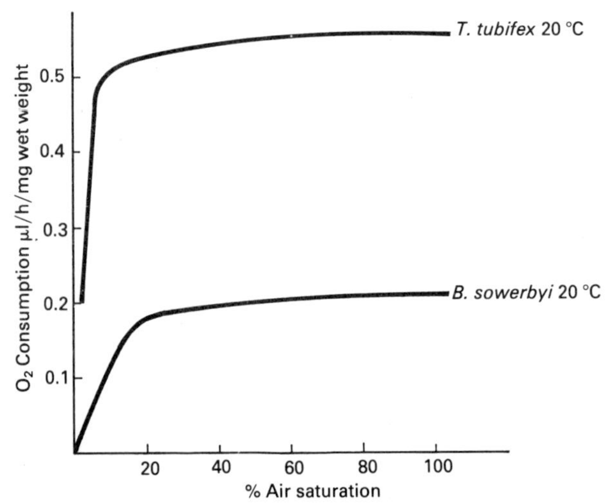

Fig. 2.13. The rate of oxygen consumption in relation to dissolved oxygen concentration, as percentage of air saturation in two species of tubificid worm, *Tubifex tubifex* (data from Palmer, 1968) and *Branchiura sowerbyi* (data of Aston, 1973) (adapted from Aston, 1973).

favoured by the group (Learner *et al*, 1978). *Nais elinguis* appears particularly tolerant of pollution (Eyres *et al*, 1978) and may increase in numbers twenty-fold below a sewage outfall (Szczesny, 1974).

As the water below a discharge of organic material becomes more oxygenated, tubificids decrease in abundance and are replaced by the midge larva (blood worm) *Chironomus riparius* (= *C. thummi*). *C. riparius* cannot withstand oxygen conditions as low as those tolerated by tubificids (Hynes, 1960). The *Chironomus* zone contains species other than *C. riparius*, including carnivorous chironomids such as the Tanypodinae, which feed on tubificids and small chironomids. Wilson and McGill (1977) have used pupal exuviae (the 'skins' which remain floating on the surface of the water after the adult midges have hatched) to characterize the distribution of chironomids in streams. Figure 2.14 shows how *C. riparius* dominates the chironomid community below a sewage outfall, being absent in the clean water immediately above. As the water self-purifies downstream the proportion of *C. riparius* decreases until it is absent 1 km below the discharge. The multivoltine life-cycle of *C. riparius*, together with the selection of sites by ovipositing females, gives the species numerous opportunities during a year to invade a site which has become suitable (Gower and Buckland, 1978) e.g. temporary pollution caused by organic run-off from a farmyard.

Chironomus, like tubificids, has haemoglobin in its blood and the content may reach up to 25 per cent of the value for human blood (Neuman, 1961). The haemoglobin has a molecular weight of 34 400, half that of mammalian blood (Svedberg and Eriksson–Quensel, 1934) and the pigment content of the blood increases when the water is poorly aerated (Fox, 1954). It acts as a carrier mainly when the oxygen tension of the water is low, at a time when the amount of oxygen required by the animal cannot be supplied by physical solution. *Chironomus riparius* lives in a tube, which is kept oxygenated by the undulatory movements of the animal's body. This extends the layer of oxygenated mud into the sediments and increases their rate of oxidation (Westlake and Edwards, 1957).

Below the chironomid zone, the isopod *Asellus aquaticus* becomes numerous (Fig. 2.7D, p. 40), especially where beds of *Cladophora* occur. Leeches, molluscs and the alder fly *Sialis lutaria* are also often abundant in this zone (Pentelow *et al*, 1938). The amphipod *Gammarus pulex* is very much more sensitive to organic pollution and Hawkes and Davies (1971) have suggested that the ratio of *Gammarus* to *Asellus* may be a useful indicator of organic enrichment. Laboratory experiments by Hawkes and Davies (1971) have shown

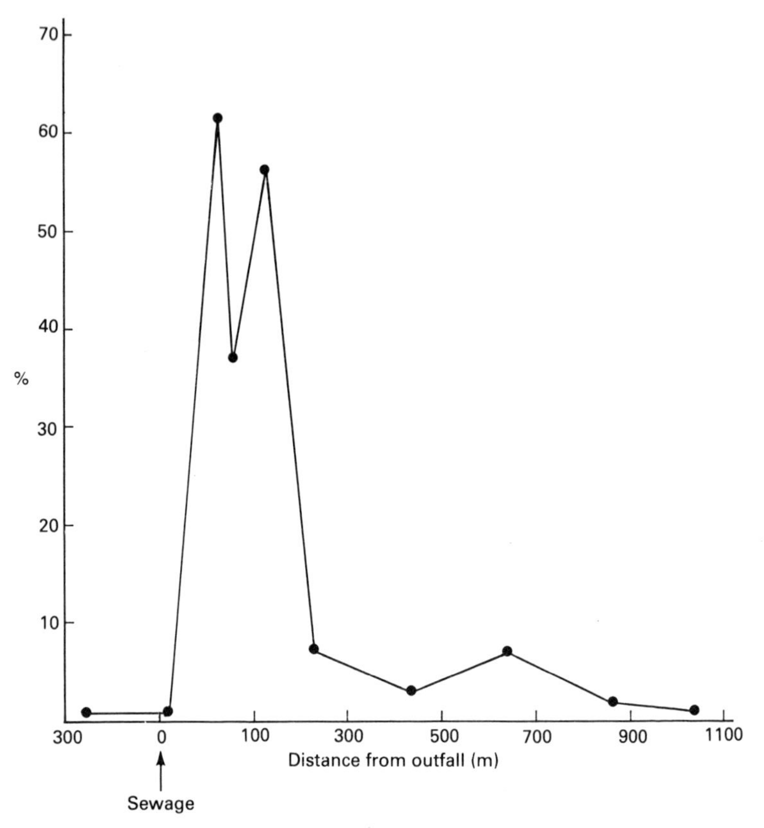

Fig. 2.14. The number of *Chironomus riparius* pupal exuviae as a percentage of the total number of chironomid exuviae at stations above and below a sewage outfall on the River Chew in south-west England (adapted from Wilson and McGill, 1977).

that *Gammarus* was killed within five hours at an oxygen concentration in the water of 1 mg $l^{-1}O_2$. Its distribution in rivers was largely determined by the number of hours during the night that the oxygen concentration fell below this critical concentration.

As the self-purification process continues, other species of invertebrates appear in the community. The most sensitive species to organic pollution are the stoneflies (Plecoptera) and to a lesser extent the mayflies (Ephemeroptera), which are often absent even at mild levels of pollution. *Amphinemura sulcicollis* is more tolerant than other stoneflies of organic pollution and *Baetis rhodani* and *Caenis horaria* are more tolerant than other mayflies (Hynes, 1960).

Fishes

Fishes are considerably more mobile than any of the organisms discussed so far and they can potentially avoid pollution incidents. Mossewitch (1961) has described how fishes move out of an area of the River Ob in Siberia when oxygen levels fall sharply in winter due to the drainage of anoxic water from surrounding marshes. Using simulated streams, Stott and Cross (1973) have show that roach (*Rutilus rutilus*) can avoid a localized area with low oxygen concentration.

The generalized relationship between the oxygen consumption of a fish and the oxygen tension of the water is shown in Fig. 2.15. The standard metabolic rate (equivalent to the basal metabolic rate) is independent of the oxygen concentration of water down to the incipient lethal level, below which survival becomes difficult. The increase in the standard rate which sometimes occurs before the incipient lethal level is reached is caused by an increased respiratory activity necessary at low oxygen tensions. The active metabolic rate is dependent on oxygen tension to some fairly high critical level (the incipient limiting level) after which it is independent of oxygen tension.

When the oxygen supply becomes deficient, a fish will breathe more rapidly and the amplitude of the respiratory movements will increase. Jones (1947) observed the opercular movements of three-spined sticklebacks (*Gasterosteus aculeatus*) in a respiration chamber

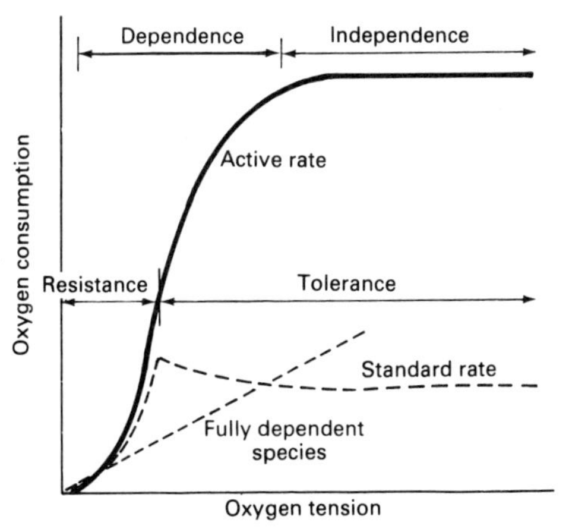

Fig. 2.15. The relationships between the active and standard rates of oxygen consumption of fishes and the oxygen tension of the water (from Warren, 1971; after Hughes, 1964).

and replaced the water with de-oxygenated water after 10 minutes. (Fig. 2.16). Aerated water was re-admitted after 28 minutes. A marked increase in opercular movements occurred initially, followed by a rapid decline. Upon re-aeration of the water the rate of opercular movement gradually returned to the original level.

Oxygen lack and carbon dioxide excess, both of which occur with organic pollution, increase the ventilation volume of fishes and, at lower levels of oxygen, the cardiac output is reduced. This reduces the rate of passage of blood through the gills, so allowing a longer period of time for uptake of oxygen, and also conserves oxygen by reducing muscular work. The zone of resistance is reached when the oxygen tension in the water is so low that the homeostatic mechanisms of the fish are no longer able to maintain the oxygen tension in the afferent blood and the standard metabolism begins to fall.

Severe organic pollution usually renders rivers fishless for considerable distances downstream of the discharge (Hynes, 1960). Fish usually begin to re-appear in the *Cladophora/Asellus* zone, the stickleback often being the first species to occur. Many pollutional and environmental factors affect the abilities of fishes to take up oxygen, so that their reactions to organic pollution in field situations are often complex.

Organic pollution tends to be most severe in the lower reaches of rivers and in estuaries and this can cause particular problems for migratory fish such as salmon (*Salmo salar*) and sea trout (*Salmo*

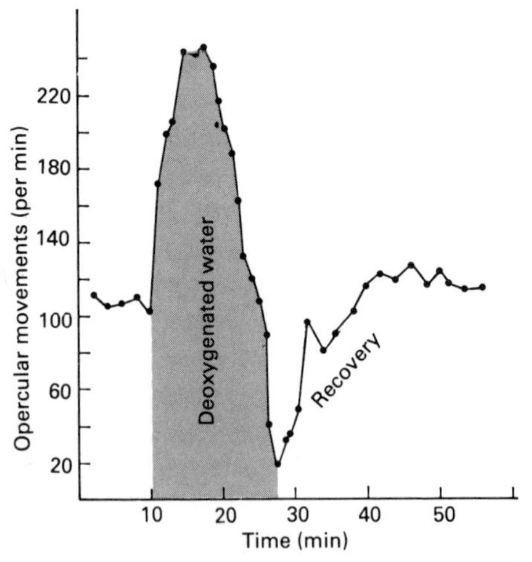

Fig. 2.16. Rate of opercular movements for a stickleback in aerated and de-oxygenated water (from Jones, 1947).

trutta), which have high oxygen requirements. These fishes may be prevented by severe pollution downstream from reaching their breeding grounds in the headwaters of rivers, even though conditions are perfectly satisfactory there.

Effects of organic pollution on the river community

The preceding section has examined the general effects of organic pollution on various groups of organisms. Some studies on the changes in community structure caused by organic pollution will now be examined. Complete studies of course are relatively rare, so this section will concentrate on invertebrates, with some information on fishes.

A study of the Tamar catchment

Nuttall and Purves (1974) surveyed the macro-invertebrate community, by taking kick samples, in the River Tamar in south-west England during 1970 and 1971. The river has a catchment area of 923 km^2 and a mean summer flow of 11.84 $m^3\ s^{-1}$, the mean winter flow being some three times greater. There are no large centres of population. Organic effluents come from a number of small sewage treatment works and from farm wastes. Milk product wastes and cooling water are the only industrial effluents and most of the pollution occurred in the upper part of the catchment, the lower reaches being fed by streams draining from moorland. The river supported a good population of fishes, including migratory trout and salmon.

A total of 82 macro-invertebrate taxa were identified from 98 samples taken at 51 sampling stations and the distribution and abundance of the major groups are shown in Fig. 2.17. The pollution in the catchment was mild (it was considered free of pollution in 1975 using mainly chemical criteria, Department of the Environment, (1978) and this was reflected in the widespread distribution of *Gammarus pulex*, whereas *Asellus* was found only at stations 5, 8 and 10, immediately downstream of discharges of organic wastes. Organic enrichment from sewage effluent and farm waste eliminated stoneflies from stations 3, 8, 9 and 27. Thirteen species of mayfly were present, most of them widely distributed and often abundant. Both *Rhithrogena* and *Ephemerella ignita*, which are very sensitive to organic pollution, were present at 40 stations. *Chironomus riparius*, tolerant of gross organic pollution, was abundant in association with sewage fungus at Station 43, a site receiving organic effluents from a

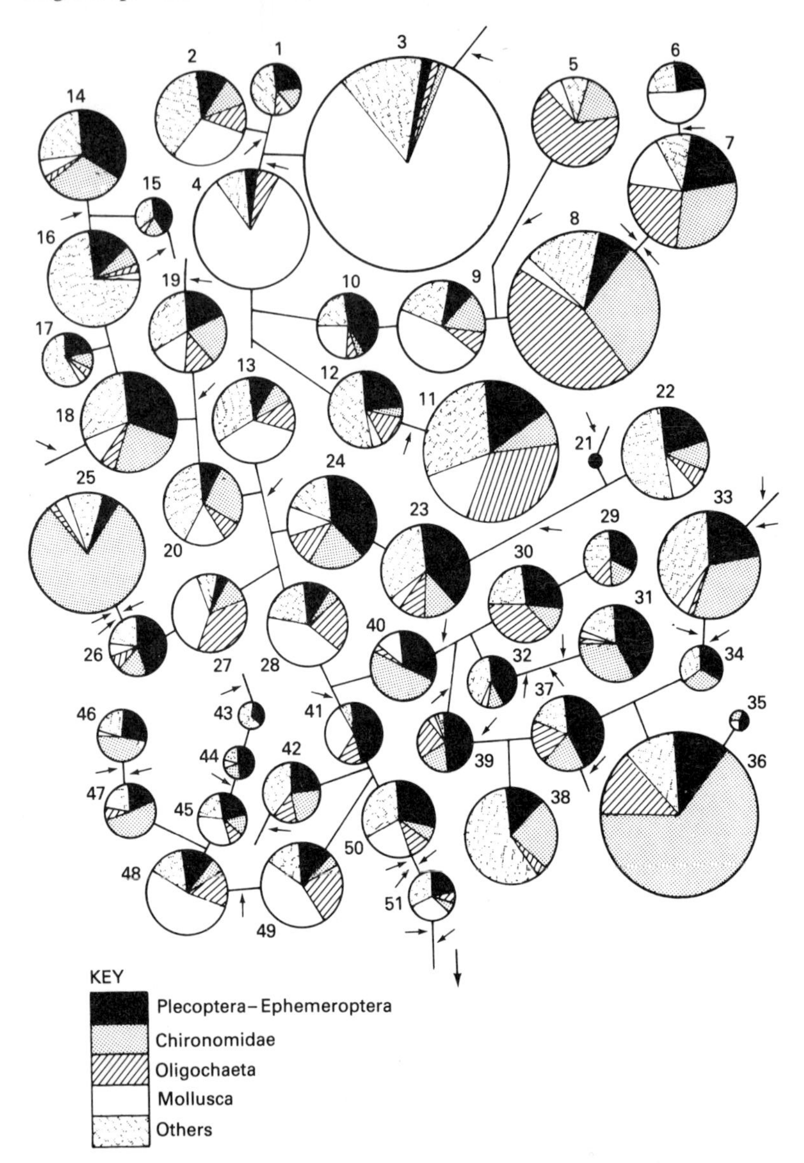

Fig. 2.17. The composition and abundance of the major macro-invertebrate groups in the Tamar catchment, with major discharges of effluents marked by arrows (adapted from Nuttall and Purves, 1974).

dairy. Similarly the snail *Potamopyrgus jenkinsi* occurred below sewage effluent and farm waste discharges.

The macro-invertebrate community in the Tamar responded to the mild pollutional conditions in the river and stoneflies and mayflies in particular, though not entirely eliminated, made up a much smaller proportion of the community where the river was receiving organic effluents, a situation favouring chironomids and snails.

A survey of the River Cynon

By contrast to the River Tamar, the River Cynon in South Wales suffers severe pollution. In 1975, using mainly chemical criteria of water quality, the bottom third of the river was classed as polluted and in urgent need of improvement, while a stretch immediately upstream of this was grossly polluted (Department of the Environment, 1978). The river received sewage and industrial effluents from a number of discharges and suspended solids from coal washery plants also occurred. In addition to organic effluents, cyanide, phenol and ammonia were also known to be present, adding to the pollutional complexity.

Learner *et al* (1971) sampled 22 stations on the river and its tributaries in 1970, 13 stations for fish and invertebrates and the remaining 9 for fish alone. The stations, together with the principal effluent discharges, are shown in Fig. 2.18. The macro-invertebrates were collected with a Surber sampler and the fishes by electrofishing.

The number of invertebrate taxa totalled 131, identification having been taken much further than in the study previously described. The distributions of selected taxa are shown in Fig. 2.19. Organic enrichment upstream of C5 affected the number of species present. In the middle and lower Cynon, the proportion of the tubificid worm *Limnodrilus hoffmeisteri* to other tubificids increased. Tubificid and enchytraeid worms became increasingly important in these lower reaches, whereas chironomids decreased in both numbers and diversity. Stoneflies were restricted to the upper stations on the river and the ten species of mayflies were largely so, though *Baetis rhodani*, fairly tolerant of pollution, extended in small numbers downstream. Most caddis and beetles were in the upper reaches of the river. Molluscs, favoured by organic enrichment, were most abundant in the middle and lower reaches.

The macro-invertebrate community changed sharply above and below C4, with an increase in the diversity of naidid worms and gastropods and a change in the species of chironomids downstream, due probably to the discharge of coal washings into the river.

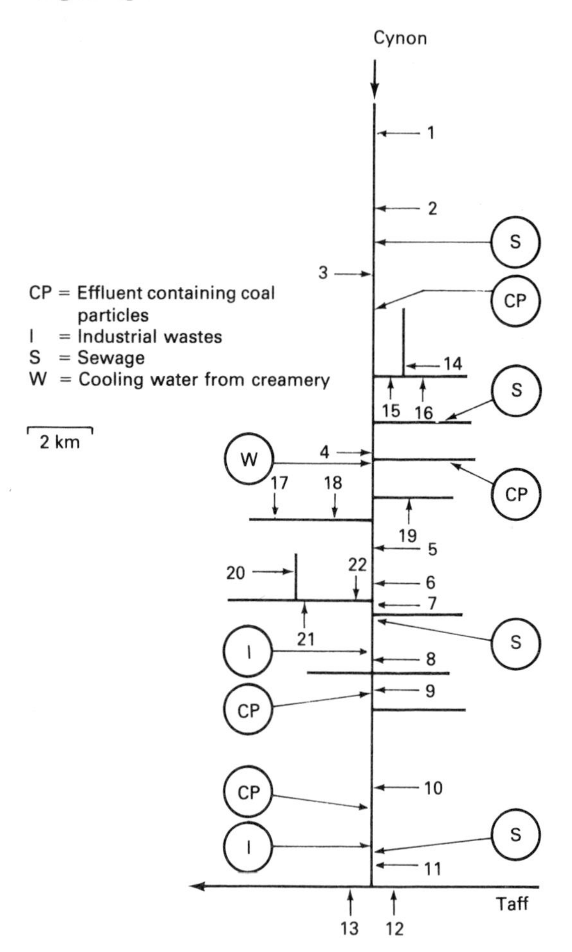

Fig. 2.18. The sampling stations, pollutional status, and sites of principal effluent discharges on the River Cynon (adapted from Learner *et al*, 1971).

The distribution of fishes in the River Cynon is shown in Fig. 2.20. Six species were present and a further two, roach and gudgeon (*Gobio gobio*), were present in the River Taff at its confluence with the Cynon, but were never recorded in the study river. Fishes were entirely eliminated between C8 and C9, where industrial discharges entered the river, and only three species subsequently entered downstream, with reduced density and biomass. Trout, bullheads (*Cottus gobio*), eels (*Anguilla anguilla*) and stone loach (*Noemacheilus barbatulus*) occurred in the Taff above its confluence with the Cynon but did not occur below. Learner *et al* (1971) considered that toxic

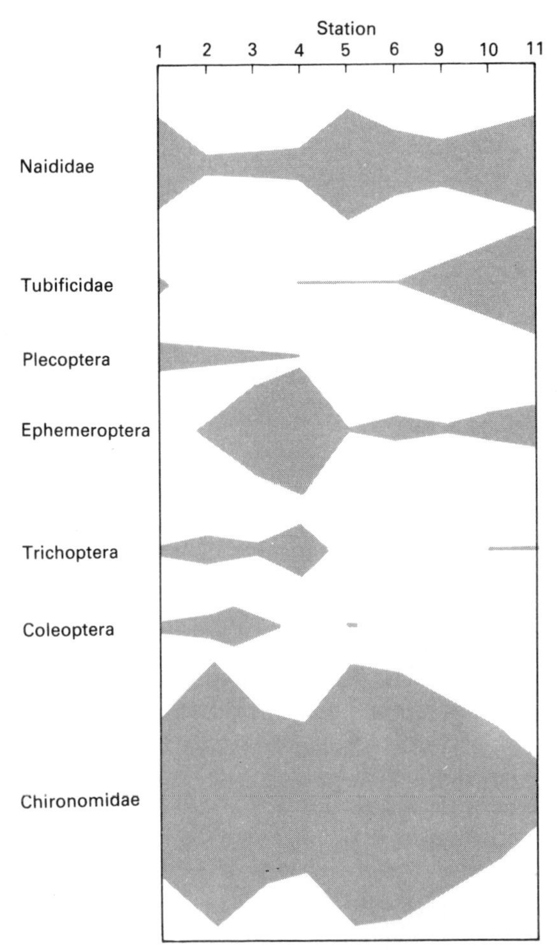

Fig. 2.19. The proportional representation of selected taxa occurring at each station on the River Cynon (adapted from Learner *et al*, 1971).

discharges, rather than organic pollution, were largely responsible for the distribution of fishes, thus emphasizing the difficulty of assigning the cause of a particular set of observations to a specific pollutant.

The effects of a single polluting incident

The River Ray, a tributary of the upper Thames in south midland England, receives secondarily treated sewage near to its source, making up a large proportion of the river's summer flow and resulting

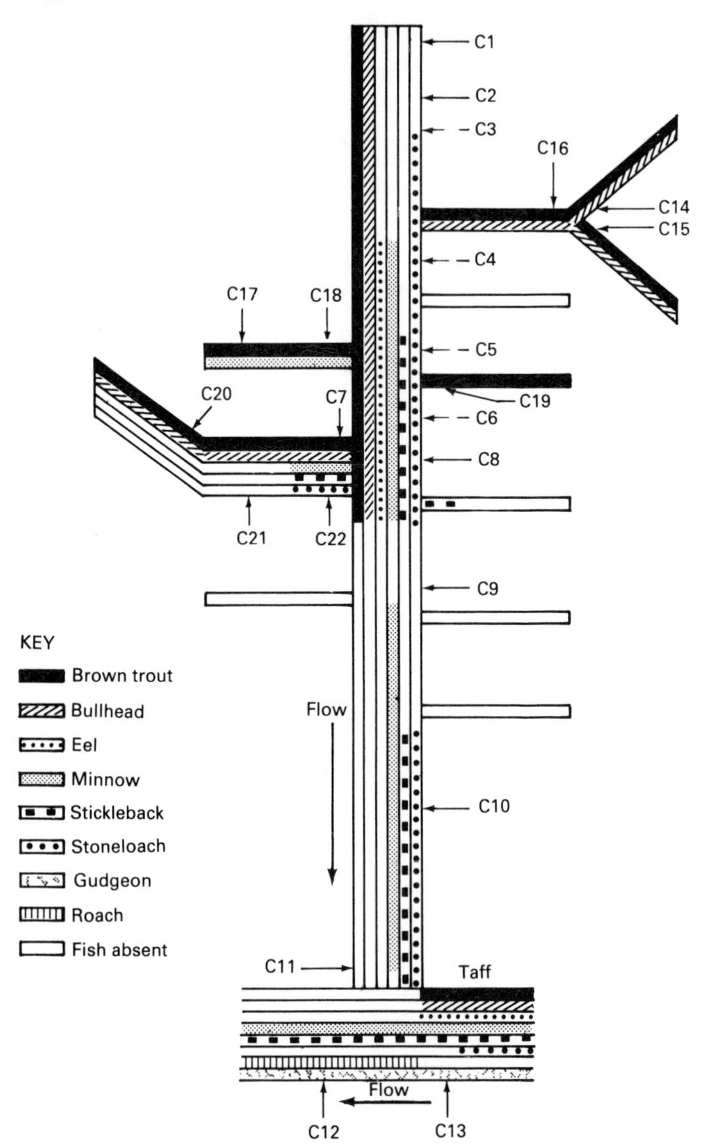

Fig. 2.20. The distribution of fishes on the River Cynon (adapted from Learner *et al*, 1971).

in a generally impoverished fauna. Hawkes (1978) has sampled three sites on the Ray and one on the Thames, below the confluence with the Ray, for a number of years. A few weeks after one sampling programme, in October 1970, an industrial dispute by sewage workers resulted in untreated sewage being released into the Ray for a period of 99 hours. Many thousands of fishes were killed in the Thames below the confluence with the Ray during this period.

A survey of the Ray after the incident showed that the aquatic plants *Cladophora* and *Ranunculus*, previously common in the river, had been completely destroyed, leaving the bed of the river devoid of plants. The fauna was restricted to tolerant species before the pollution incident but, after the release of crude sewage, only tubificid worms and *Asellus aquaticus* survived in the lower reaches of the Ray and their populations had been decimated (Fig. 2.21). The effect became more noticeable further downstream of the discharge, so de-oxygenation, resulting from the breakdown of sewage by microorganisms, rather than toxic effects were the obvious cause of the loss of invertebrates from the river. In the Thames, the relatively high diversity of invertebrates was severely reduced and their populations fell ten-fold, due to the raw sewage.

A series of surveys during the following year showed that, within twelve months, the macrofauna in the Ray and the Thames had largely returned to the pre-pollution level.

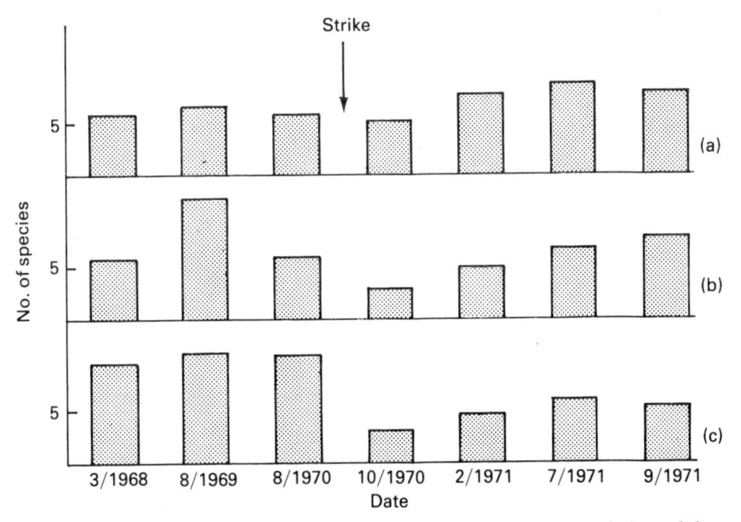

Fig. 2.21. The effects of a pollution incident on the River Ray (adapted from Hawkes, 1978). a) 2.1 km, b) 4 km, c) 10.8 km below discharge.

Recovery from a complex pollutional condition

The River Ray provides an example of the effects of a relatively simple polluting incident on the river fauna and its subsequent recovery. The pollution entering the Cynon was complex. The lower reaches of rivers and estuaries receive pollution from many sources and the amelioration of conditions may require improvements to a number of discharges. The upper Clyde estuary, receiving water from the large industrial conurbation around Glasgow, will provide an example. While not strictly freshwater, because of the tidal conditions, the upper estuary nevertheless supports a number of freshwater organisms which can tolerate saline conditions.

The Clyde estuary has been grossly polluted at least since Victorian times. The first sewage treatment works was commissioned in 1894, providing primary treatment, followed by a number of others. The settled sewage sludge was disposed of at sea from 1910. Secondary treatment came into operation at one sewage works from 1968 and a new works providing secondary treatment was completed in 1974, while a further plant is being constructed. In addition, certain industries, previously supplying a polluting load, have closed down. There has been a steady improvement in the dissolved oxygen content in the estuary.

The freshwater reaches of the river, immediately above the tidal weir, had a sparse population of four oligochaetes, a leech and *Asellus* in 1968, before biological treatment had begun (Mackay *et al*, 1978). By 1973 the snails *Physa fontinalis* and *Lymnaea peregra*, the bivalve *Pisidium* and the triclad *Dendrocoelum lacteum* had been added to the fauna.

The situation in the upper estuary at seven stations downstream of the tidal weir is shown in Fig. 2.22. In the early years there was a complete absence of macrofauna in the first 10 km downstream of the weir, while the next 10 km had only oligochaetes. Further downstream the typically estuarine polychaetes *Nereis diversicolor* and *Pygospio elegans* were added. From 1972 the fauna began to respond to the decrease in polluting load, with the first oligochaetes being recorded immediately below the tidal weir in 1972 and three species having become established by 1974. By 1976 several species of oligochaetes had become established at all sites, while a number of species of polychaetes had extended their range up the estuary. Crustaceans and molluscs were also occasionally recorded.

Biological surveys have established that the Clyde estuary is improving in quality, with a reduction in the polluting load from a variety of sources, though the upper estuary still suffers de-oxygenation in hot,

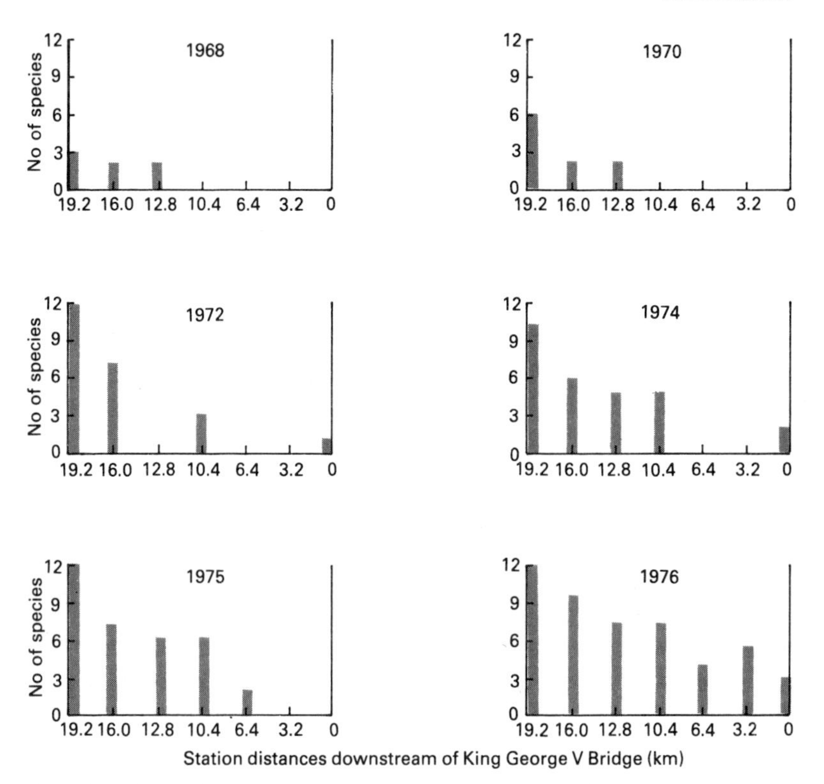

Fig. 2.22. The number of macro-invertebrate species recorded at stations in the upper Clyde estuary, 1968–76 (adapted from Mackay *et al*, 1978).

dry weather during the summer. It is hoped that further improvement will allow salmon and migratory trout to negotiate the estuary to their breeding grounds in the headwaters of the Clyde catchment, where they have not been recorded for many years.

Conclusions

The absence of a particular species or group from a river may not be indicative of pollution because not all reaches are suitable for all invertebrates. Stoneflies, for example, are largely confined to eroding substrata with fast currents, and do not occur in the slow-moving, silty, lowland reaches of rivers, even where these are free of pollution. Various zones can be recognized along a river and these have characteristic animals and plants associated with them (Huet, 1954; Hawkes, 1975).

Jones and Peters (1977) have examined the distribution of invertebrates in relation to river flow and have isolated groups of invertebrates which were characteristic of specific flow regimes. They studied 80 sampling sites on 43 unpolluted rivers in Britain. Ninety-three species out of a total of 439 showed significant association at the 95 per cent confidence level, using Fager's (1957) index of affinity. These significant associations were then subjected to a recurrent groups analysis (Fager, 1957) in which groups where all taxa were significantly associated with one another were isolated. Jones and Peters found eight such groups which could be related to the flow regimes of rivers. For example, stoneflies were restricted to groups 5 and 6, both of which occurred in upland, spatey rivers with surface velocities greater than 0.4 m s^{-1}. Group 2 occurred in slow-flowing, lowland rivers with a surface velocity of less than 0.4 m s^{-1} and consisted of three leeches, *Asellus*, the beetle *Elmis* and the mayfly *Baetis rhodani*. Group 7 consisted of three mayfly species which occurred in swiftly flowing rivers supporting considerable macrophyte growth. Overall, four of the groups were found statistically to be confined to a particular range of flow conditions, while three groups occurred over the entire range of river flow types.

The work of Jones and Peters (1977) is one of the few attempts to establish the natural factors governing the structure of macroinvertebrate communities. Similar analyses are needed for a range of physical and chemical parameters. Once the natural community of invertebrates and other organisms for a particular stretch of water can be predicted, deviations due to organic and other pollution can be more readily assessed. Species will also be affected by biotic changes caused by pollution, such as an increased food supply, or the elimination of competitors or predators. For a complete understanding of the action of pollution we will need to disentangle the interwoven effects of a range of pollutants acting simultaneously at the physiological, population and community level.

Chapter 3

EUTROPHICATION

Introduction

Over the last two decades, with the widespread installation of sewage treatment plants, the organic pollution of freshwaters has generally become less of a problem. During the same period concern over pollution by inorganic substances has grown and is now a central issue in the control of water pollution.

The terms *oligotrophic* and *eutrophic* were originally introduced by Weber in 1907 to describe nutrient conditions in the development of peat bogs. Naumann (1919) introduced the terms to the limnological world, classifying lakes as having oligotrophic water if they were clear in summer and eutrophic water if they were turbid due to the presence of algae. Considerable confusion over these terms has developed as other definitions have arisen. In particular, oligotrophic lakes have been defined as having a low productivity and eutrophic lakes as having a high productivity. Whilst in general terms high nutrient levels and high productivity go together this may not always be the case. For instance, marl lakes have highly calcareous, nutrient rich water (eutrophic), with a heavy precipitation of carbonates. Under these conditions phosphates and many micro-nutrients form very insoluble compounds which precipitate to the bottom of the lake. Thus algal productivity is low and Wetzel (1968) has described marl lakes as oligotrophic.

Eutrophication will be defined as *an increase in the rate of income of nutrients* (Edmondson, 1974). The income of nutrients may be considered as *artificial* or *cultural* eutrophication if the increase is due to human activities, or *natural* eutrophication if the rate of increase is caused by a non-human process, such as a forest fire. From our definition of pollution in Chapter 1 (p. 1) only artificial eutrophication will concern us here.

Table 3.1. The general characteristics of oligotrophic and eutrophic lakes.

	Oligotrophic	*Eutrophic*
Depth	Deeper	Shallower
Summer oxygen in hypolimneon	Present	Absent
Algae	High species diversity, with low density and productivity, often dominated by Chlorophyceae	Low species diversity with high density and productivity, often dominated by Cyanophyceae
Blooms	Rare	Frequent
Plant nutrient flux	Low	High
Animal production	Low	High
Fish	Salmonids (e.g. trout, char) and coregonids (whitefish) often dominant	Coarse fish (e.g. perch, roach, carp) often dominant

The general characteristics of oligotrophic and eutrophic waters are given in Table 3.1. A third general category of lake includes those which receive large amounts of organic matter from terrestrial plant origin – *dystrophic* or brown-water lakes, because of the heavily stained water. They usually have a low planktonic productivity.

It is often considered that all lakes are originally oligotrophic and become eutrophic as they age. However, as Edmondson (1974) has pointed out, a lake may become more productive with time, but this is due to the effect of shallowing, which affects the way a lake converts nutrients into organisms, rather than an increased input of nutrients.

The sources of nutrients

A number of compounds and elements (e.g. silicon, manganese, vitamins) may at times be limiting to algal growth but it is only excesses of nitrogen and phosphorus which give rise to algal nuisances. In the majority of *lakes* phosphorus is normally the limiting element because the amount of biologically available phosphorus is small in relation to the quantity required for algal growth, so that an increase in phosphorus will result in an increase in productivity. If nitrogen becomes limiting some blue-green algae are able to fix nitrogen and grow provided phosphorus is not limiting. Evidence will be provided for this later in the chapter. In marine waters it is usually nitrogen which is limiting.

The majority of polluting nutrients enter watercourses and lakes in effluents from sewage treatment works, in untreated sewage or from farming activities. Sources might be *discrete*, such as a specific sewage outfall, or *diffuse*, such as from farmland within the catchment. The nutrient flows within the human food web are shown in Figure 3.1.

In some cases, of course, eutrophication which proves undesirable from man's viewpoint may have its cause in mainly natural events. An example is described in the work of Moss (1978) on Hickling Broad, Norfolk, England. Moss analysed a sediment core from the bottom of the lake and found an increase in the rate of organic sedimentation during the 1930s, which he associated with the fertilization of agricultural lands. However, there was no increase in the phytoplankton population. During the 1960s there was a marked increase in epiphytic diatoms and phytoplankton and a loss in the

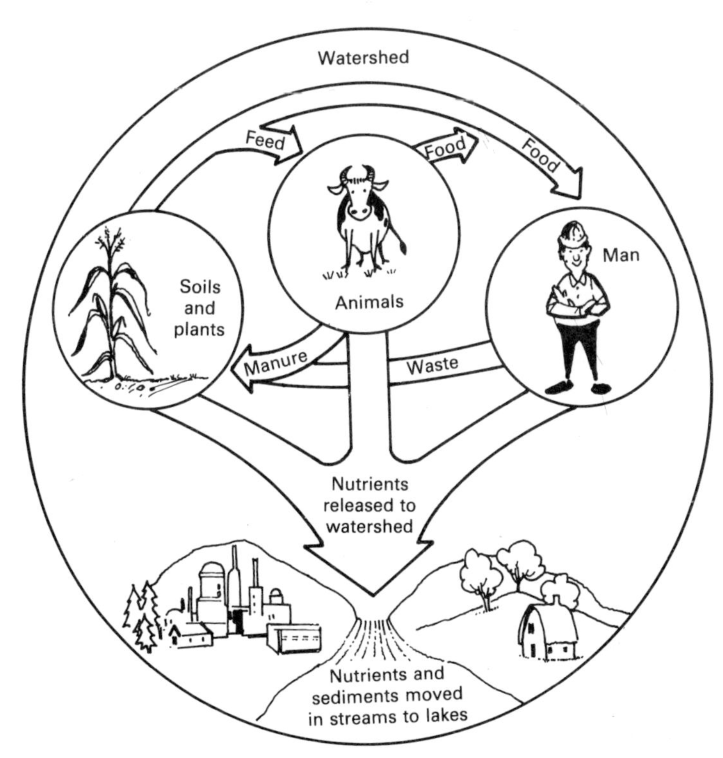

Fig. 3.1. Nutrient flows within the human food web and from its components to the aquatic environment (from Porter, 1975).

Table 3.2. The quantities of nitrogen and phosphorus derived from various types of land use in the USA (from Lee, Rast and Jones, 1978).

Land use	*Total phosphorus g m^{-2} y^{-1}*	*Total nitrogen g m^{-2} y^{-1}*
Urban	0.1	0.5
Rural/agriculture	0.05	0.5
Forest	0.01	0.3
Other sources:		
rainfall	0.02	0.8
dry fallout	0.08	1.6

dense and varied macrophyte growth. This sudden change was attributed by Moss to an increase in the number of Black-headed Gulls (*Larus ridibundus*) roosting on the lake. The roost appears to have increased ten-fold over two decades, with a winter maximum estimated at some 250 000 birds in the mid-1970s.

Table 3.2 gives a general summary of the quantities of nitrogen and phosphorus which are derived from different types of land use in the United States. Note the considerable increase in the export of phosphate as forest is converted to agricultural use and agricultural land is subject to urban development. The relative importance of sources of nutrients will vary, of course, with the type of land use in the catchment.

Urban sources of nutrients

Nutrients from urban sources may be derived from domestic sewage, industrial wastes and storm drainage. The contribution of nitrogen and phosphorus per person averages 10.8 g N/capita/day and 2.18 g P/capita/day, though there is a considerable range (Vollenweider, 1968). Detergents containing phosphates, first developed in the 1940s, have become a very important source of phosphorus in domestic sewage. Figure 3.2 shows the increase in usage of detergents in the United States, Japan and Britain. Phosphorus from detergents made up 47–65 per cent of the total phosphorus in sewage from six English sewage works in 1971, compared with 10–20 per cent in 1957 (Devey and Harkness, 1973).

Industrial sources of nutrients may be locally important, depending on the type of industry, the volume of effluent and the amount of treatment it receives. For instance, the brewing industry which released some 10 680 m^3 effluent each day into rivers in England and

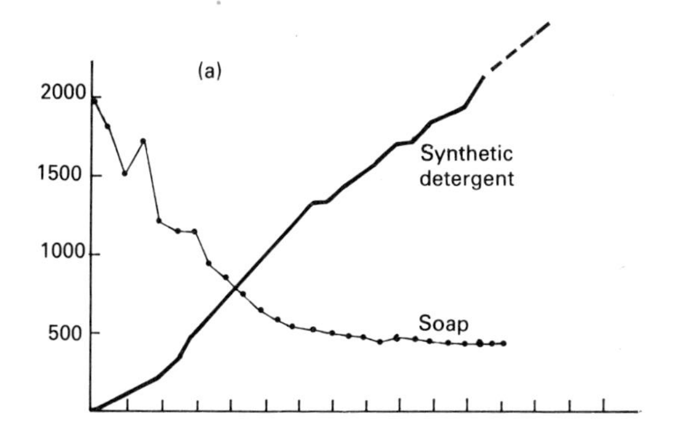

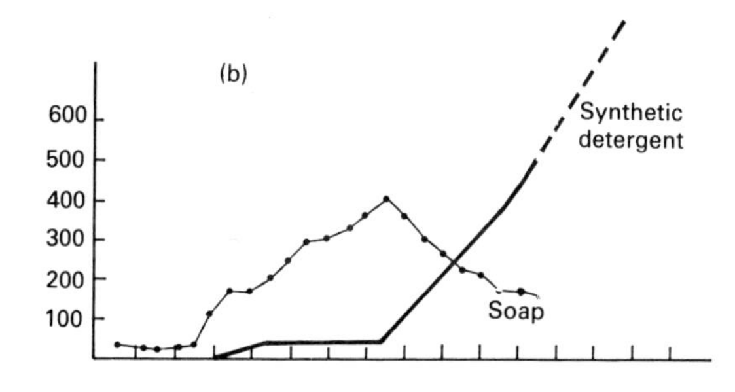

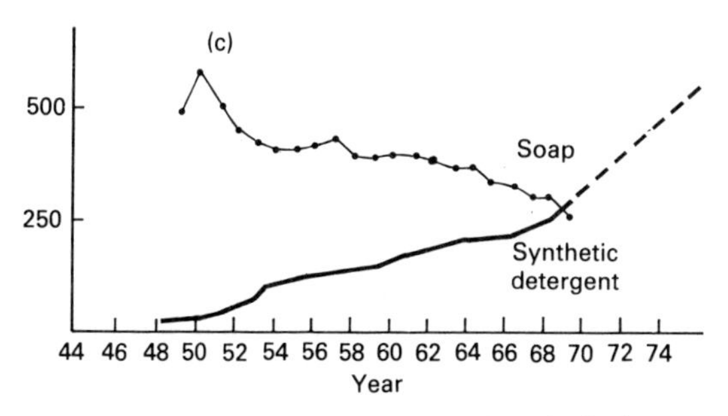

Fig. 3.2. Trends in the consumption of synthetic detergents and soaps (metric tons × 10^3) for (a) USA (b) Japan (c) United Kingdom (after Devey and Harkness, 1973).

Wales in 1975 (Department of the Environment, 1978), produces an effluent containing some 156 mg l^{-1} N and 20 mg l^{-1} P (Vollenweider, 1968). Food processing generally and those concerns requiring substantial washing procedures, e.g. the wollen industry, are likely to produce effluents containing high concentrations of nitrogen and phosphorus.

Rural sources of nutrients

Rural sources of nutrients include those from agriculture, from forest management and from rural dwellings, of which the first is the most universally important. Rural dwellings tend to dispose of their sewage into septic tanks which might cause local pollution, though Lee *et al* (1978) consider them generally unimportant. However, summer houses, which often have primitive sewage disposal facilities, are often built on lakesides.

Nutrients are lost from farmland in three ways (Tomlinson, 1971):

1. by drainage water percolating through the soil leaching soluble plant nutrients;
2. by inefficient return to the land of the excreta of stock;
3. by the erosion of surface soils or by the movement of fine soil particles into subsoil drainage systems.

The use of fertilizers has vastly increased during this century and in the USA the total inorganic fertilizer use is equivalent to 40 kg/person/year (Porter, 1975).

Nitrogen and phosphorus behave differently in soils. The nitrate anion is fairly mobile because of the predominently negative charge on soil particles so that it is readily leached if it is not taken up by plants. By contrast, phosphate is precipitated as insoluble iron, calcium or aluminium phosphate and then is released only slowly.

The solubility of nitrate means that agriculture is a major contributor to nitrate loadings in freshwaters. Owens (1970) considered that agriculture accounted for 71 per cent of the mass flow of nitrogen in the River Great Ouse in the English Midlands, compared with only 6 per cent for phosphorus, the balance being from sewage effluents. Some 50 per cent of 200–250 kg N ha^{-1} fertilizer used on potato fields in eastern USA was estimated to be lost to groundwater (Porter, 1975).

The concentration of nitrate in rivers follows closely that of river flow (Fig. 3.3). Levels are low during the summer, even when ferti-

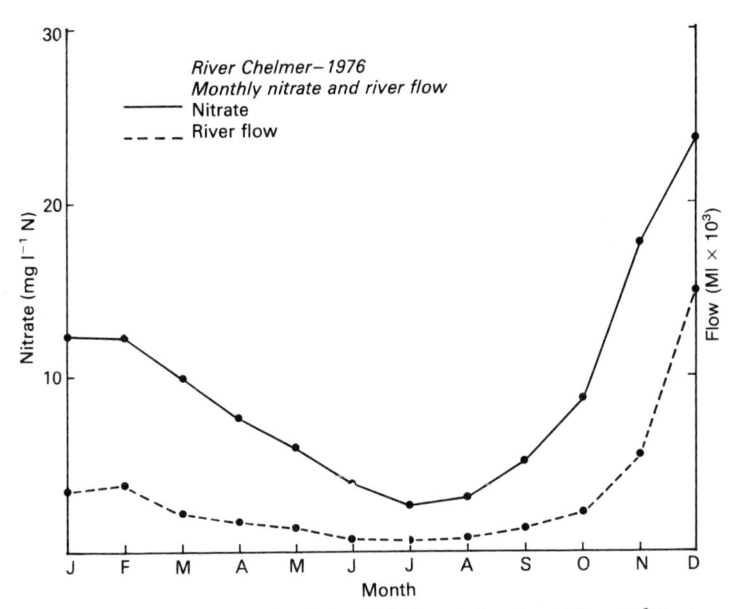

Fig. 3.3 The monthly nitrate levels (mg l^{-1} N) and river flow (Ml × 10^3) in the River Chelmer, eastern England, 1976 (from Slack, 1977).

lizer is being added, because the growing plants utilize nitrogen as soon as it becomes available. There is also little net downward movement of water in the soil during the summer because of high rates of evaporation and transpiration. With a decrease in transpiration and evaporation in autumn and winter, nitrate is leached from the soil and levels in rivers rise. The rate of loss declines again in late winter because soluble nitrate reserves are depleted and low temperatures reduce the rate of nitrification. The annual nitrate level in rivers closely follows the annual levels of fertilizer application within the catchment (Fig. 3.4).

The loss of phosphate by leaching from agricultural land is negligible, so that the input to freshwaters is largely by erosion. Arable farming increases the natural rate of erosion of land because the soil is bare over many of the winter months. Holt *et al* (1970) have reported losses of phosphorus in water of up to 6 kg ha^{-1} $year^{-1}$, which was equivalent to 60 per cent of the fertilizer applied to the land. Much of this phosphorus is tightly bound to the soil particles and may not be biologically available when it reaches freshwaters. However, Cooke and Williams (1973) have suggested that the solubility of phosphate is enhanced when the soil is deposited as mud,

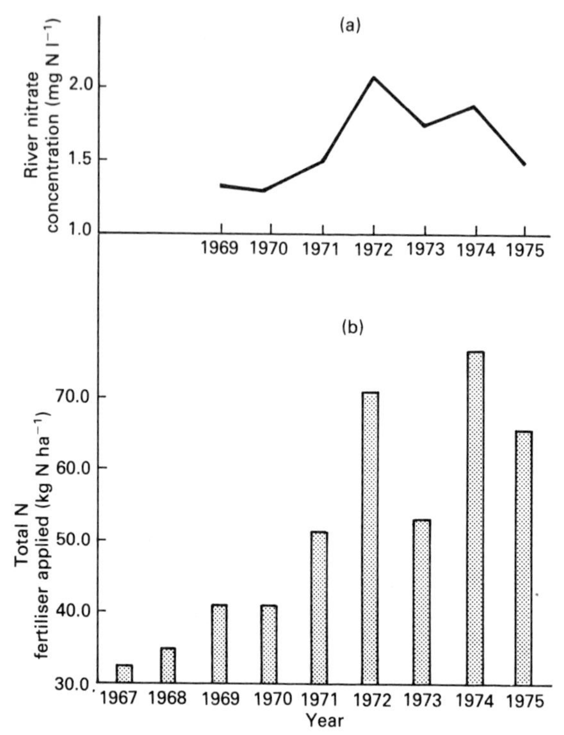

Fig. 3.4. The mean concentration of nitrate nitrogen in rivers entering Lough Neagh (a) and the mean total nitrogen fertilizer applied to land (b) in Northern Ireland (from Smith, 1977).

which then remains permanently submerged and may become anaerobic.

The other chief source of nutrients from the agricultural industry is derived from animal farming, especially where this involves intensive rearing. Labour costs for handling farmyard manure are high and areas of high livestock rearing are often distant from areas of intensive arable farming, where the wastes could be used, so that there are difficulties in disposing of manure. Farmyard manure has always been a potential source of pollution, but the problem has become more acute because although there are now fewer farm units, these are producing more livestock. This is illustrated for the chicken industry in the United States in Fig. 3.5. The number of farms with chickens has fallen dramatically since 1950, but there has been a great increase in the size of the flocks. Similar trends are apparent in beef and pig rearing. British livestock excretes some 175 000 tonnes of

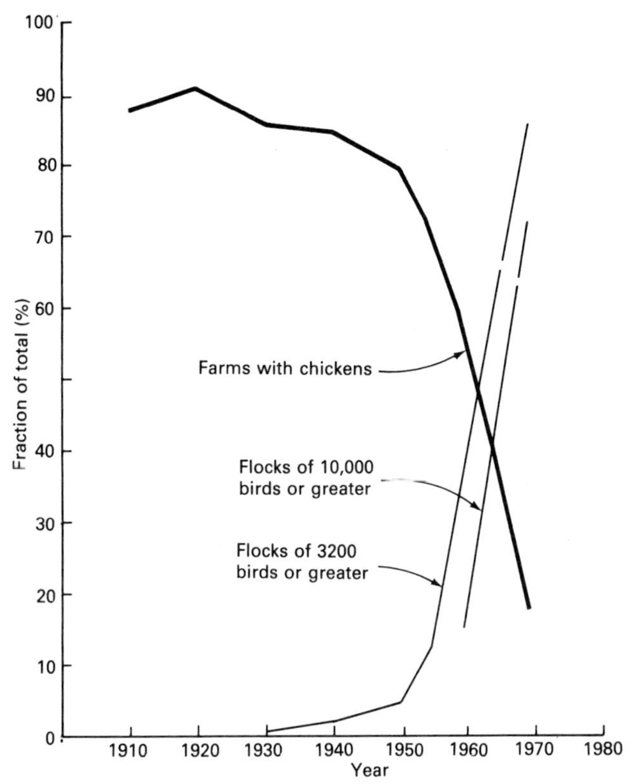

Fig. 3.5. Changes with time in the number of farms with chickens and the sizes of flocks in USA (from Porter, 1975, after Jewell *et al* 1974).

phosphorus each year, compared with 45 000 tonnes from the human population (Cooke and Williams, 1973). On average, about two thirds of manure is spread by animals living in the open, the remainder is voided indoors and presents potential disposal problems.

The management of forests may have local effects on the nutrient loading of rivers. The experiments in the Hubbard Brook Watershed in the USA, where a forest was cut and left on site, with regrowth prevented by the application of herbicides, showed that nitrate increased some fifty times compared with uncut controls (Likens *et al*, 1970). Most forest practices, of course, are nowhere near as extreme as this. In some countries of the world, forests are regularly fertilized and this may result in local eutrophication. Phosphate is added to newly established conifer plantations in Britain and with the increase in afforestation this could result in an overall increase in productivity in upland catchments.

Two case histories

Before dealing with the general effects of eutrophication I want to describe conditions in two lakes which have been studied in detail: Lake Washington in North America and Lough Neagh in the British Isles.

Lake Washington

Lake Washington (Fig. 3.6), in the north-western United States, has a surface area of 87.6 km^2 and a maximum depth of 76.5 m. Water, which is low in nutrients, enters the lake via the Cedar River from the south and the outflow is the Sammamish River, which flows into Lake Sammamish. Lake Washington is connected to the Pacific Ocean at Puget Sound, but a series of locks hold the lake level higher. The city of Seattle (population 489 000) lies between Puget Sound and the western shore of the lake. The events in Lake Washington have been detailed in a series of papers by Edmondson (e.g. 1969, 1970, 1971, 1972a,b 1979).

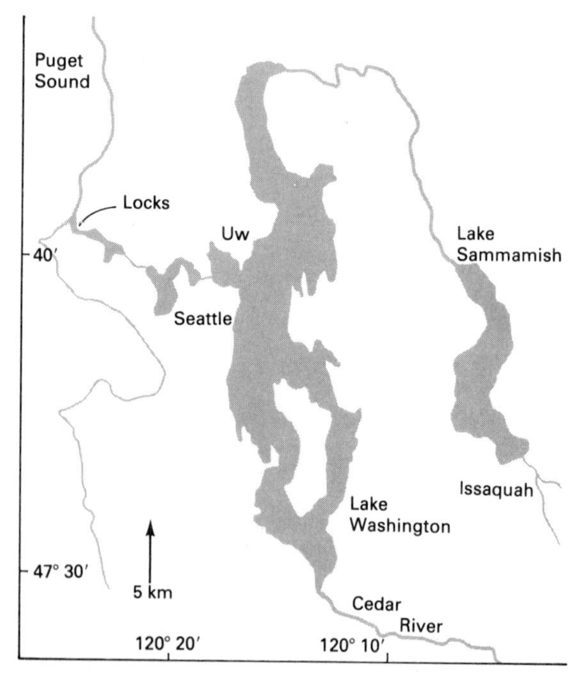

Fig. 3.6. Lakes Washington and Sammamish, USA (from Edmondson, 1969).

Early in this century raw sewage was disposed of directly into Lake Washington and, by 1926, a population of 50 000 was adding raw sewage to the lake through thirty outfalls (Edmondson, 1972b). By 1936 a series of interceptors and tunnels had been constructed to divert Seattle's sewage to the sea at Puget Sound and this reduced pollution within the lake. However, Seattle began to expand along the edges of the lake and development took place on the east side. A series of works were built to cope with the expansion and discharged secondarily treated sewage into the lake, there being ten sewage works by 1954 (Edmondson, 1972b). In addition, feeder streams into the lake were being contaminated with drainage from septic tanks. Sewage was responsible for 56 per cent of the total input of phosphorus to the lake.

Work during 1950 showed that conditions in Lake Washington were very different from those revealed in an earlier survey in 1933 and during the next few years conditions deteriorated sufficiently to attract considerable public attention to the problems. There was a greater summer oxygen deficit in the hypolimneon in 1950 compared with 1933 while the winter phosphate concentration, which gives a good index of supply to the plankton, had shown a marked increase. The summer densities of phytoplankton increased several fold during the 1950s and in 1955 a dense bloom of the blue-green alga *Oscillatoria rubescens* developed (Edmondson *et al*, 1956). A bloom can be defined as an aggregation of plankton sufficiently dense to be readily visible (Palmer, 1959). Lakeside bathing beaches were periodically closed due to pollution, but the situation was far worse in Puget Sound, which was receiving Seattle's untreated sewage.

The amenity value of Lake Washington, close to a large urban centre, had seriously declined and this gave rise to a vociferous movement to prevent further deterioration. It was decided to divert the majority of sewage from the lake to Puget Sound and at the same time to improve the quality of the effluent entering the Sound to reduce pollution there. Diversion began in March 1963, with one third of the sewage being transferred to Puget Sound. Ninety-nine per cent of the sewage had been diverted by March 1967 and the project was completed a year later at a total cost of 125 million dollars.

Lake Washington responded quickly to the reduction in nutrients. The winter levels of phosphate began to decline rapidly after 1965. Levels of nitrogen declined more slowly because agriculture is a major contributor of nitrogen to the lake. The levels of chlorophyll-*a* had fallen by 1970 to one fifth the levels in 1963 and there were corresponding increases in the transparency of the water (Fig. 3.7).

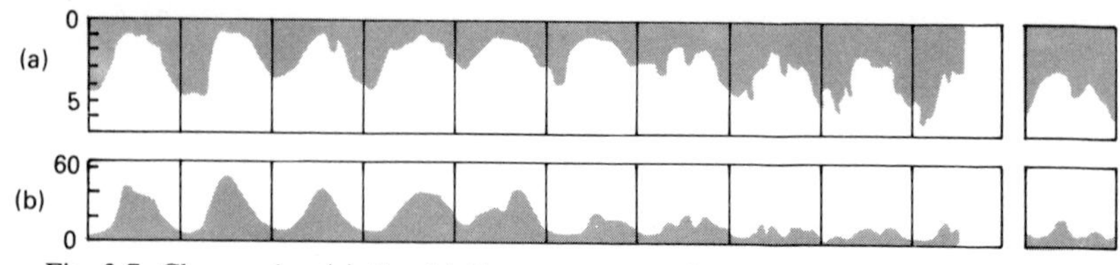

Fig. 3.7. Changes in (a) Secchi disc transparency (m, plotted with zero at top) and (b) chlorophyll in surface water ($\mu g\ l^{-1}$) in Lake Washington 1962–1971, with data for one earlier year (1950) to suggest the possible end point of the changes (adapted from Edmondson, 1972a).

The observations on Lake Washington illustrate clearly the relationship between an increase in nutrients and an increase in undesirable biological productivity, but they also show that eutrophication can be reversed if a major source of nutrients is removed.

Lough Neagh

Lough Neagh (Fig. 3.8), in Northern Ireland, has a surface area of 383 km^2 and a mean depth of 8.6 m, so it is much larger, but much shallower, than Lake Washington. It is fed by six major rivers and the Lower River Bann drains the lough to the sea. Lough Neagh is an important source of water for Ulster, there is a commercial fishery and extensive recreational use.

In 1967 a bloom of the blue-green alga *Anabaena flos-aquae* disrupted the treatment process in the waterworks (Wood and Gibson, 1973) and this encouraged a detailed study of the lake. The lake was shown to be highly eutrophic. The nitrate and phosphate levels in Kinnego Bay in Lough Neagh were some three times higher than the maximum recorded from Lake Washington, while the level of chlorophyll-*a* was some eight times higher. In the main body of the lough the values were somewhat lower than this, though it is still one of the most productive lakes in the world.

A study of a sediment core has shown that the diversity of the diatoms has decreased and the dominants have changed with time, while there has been a rapid increase in the overall production of diatoms during the last fifty years (Wood and Gibson, 1973). Blue-greens, such as *Aphanizomenon flos-aquae*, are now abundant in the lough, though they were not recorded in a survey in the early part of the century (Dakin and Latarche, 1913).

After a period of high phytoplankton growth in spring, levels of phosphate in the water become undetectable, whereas nitrate can

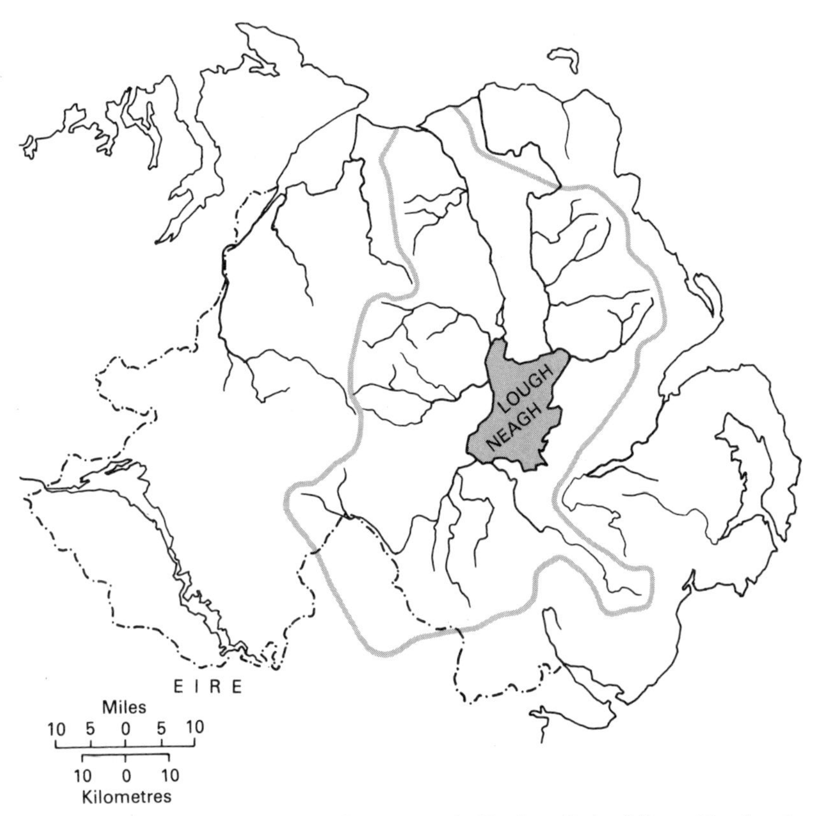

Fig. 3.8. Lough Neagh and its catchment area in Northern Ireland (from Wood and Gibson, 1973).

always be measured (Wood and Gibson, 1973), Phosphate is therefore limiting to algal growth.

The contributions of agricultural and domestic sources of nutrients to Lough Neagh have been investigated by Smith (1977). The predicted contribution of various sources of available phosphate are given in Table 3.3. If the contribution from each catchment to the lough is calculated separately there is a good relationship between the annual orthophosphate loading and the urban population density in the catchment (Fig. 3.9). There was no correlation between the usage of phosphorus fertilizer in the catchment and the mean annual concentration of phosphorus in the rivers, but there was a good relationship between nitrogen fertilizer usage and mean levels of nitrogen in the rivers.

Table 3.3. The predicted available phosphorus to Lough Neagh (after Smith, 1977).

	Tonnes P y^{-1}	*% contribution*
Land drainage	62.3	20
Rural population (unsewered)	60.2	18
Sewage disposal works serving towns with populations greater than 2000	170.9	53
Minor sewage disposal works	28.7	9
Total	322.1	100

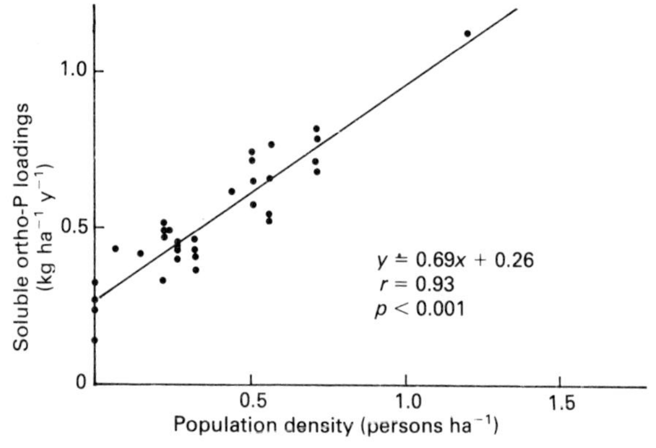

Fig. 3.9. The regression of the soluble orthophosphate loadings of the main rivers and subcatchments of the River Main, Northern Ireland (1971–1974) with the urban human population densities in the catchments (from Smith, 1977).

The timing of the input of nutrients is important because soluble phosphate will have a greater effect if it enters the lake during the spring when the growth of phytoplankton is greatest. Smith (1977) has shown that 85 per cent of the available phosphorus entering Loch Neagh during the period April to August was derived from sewage effluents.

Smith has calculated that the removal of phosphorus at those sewage works serving populations of greater than 2000 people will reduce the phosphorus available to algae in the lough by 50 per cent.

The general effects of eutrophication

We have examined two enriched lakes in some detail and can now generalize about the effects on waterbodies of an increase in the rate of nutrient income. The effects of eutrophication on the receiving ecosystem and the particular problems these cause to man are summarized in Table 3.4.

Eutrophication causes marked changes in the biota. The Norfolk Broads area of eastern England can serve as an example. The Norfolk Broads are a group of relatively small, shallow lakes formed during medieval times by the flooding of peat diggings. They are calcareous and naturally eutrophic, but the water used to be generally clear and dominated by a rich flora of Charophytes and aquatic angiosperms. The plants support a diverse fauna of invertebrates, some of them rare in Britain. The area as a whole is outstanding for its wildlife interest, but at the same time maintains a large tourist industry, based mainly on boating holidays and angling.

Conditions began to deteriorate in the Broads during the 1960s, but the area was not subject to any scientific study. A survey in 1972 and 1973 of 28 broads showed that 11 were completely devoid of aquatic macrophytes or had only a poor macrophyte growth, consisting mainly of floating-leaved water lilies (Mason and Bryant, 1975). This change is clearly illustrated in the photographs of one nature

Table 3.4. The effects of eutrophication on the receiving ecosystem and the problems to man associated with these effects.

Effects

1. Species diversity decreases and the dominant biota change.
2. Plant and animal biomass increases.
3. Turbidity increases.
4. Rate of sedimentation increases, shortening the life-span of the lake.
5. Anoxic conditions may develop.

Problems

1. Treatment of potable water may be difficult and the supply may have an unacceptable taste or odour.
2. The water may be injurious to health.
3. The amenity value of the water may decrease.
4. Increased vegetation may impede water flow and navigation.
5. Commercially important species (such as salmonids and coregonids) may disappear.

reserve taken in 1930 and in 1975 (Fig. 3.10). At the same time permanent algal blooms had developed and the transparency of the water in summer at one site, Barton Broad, measured with a Secchi disc was only 11 cm. Broads with no submerged macrophytes had a poorly developed benthic fauna dominated by tubificid worms and chironomids. A detailed study of the benthos of an unpolluted broad and a culturally enriched broad (Mason, 1977a) showed that the unpolluted site (Upton Broad) had 40 taxa in the benthos, 17 of them occurring commonly. The enriched site (Alderfen Broad) had 22 taxa, only 7 of them occurring commonly.

In the Norfolk Broads there has been a change from a macrophyte dominated community to a phytoplankton dominated community. The macrophytes have disappeared due to shading from the phytoplankton and due to the dense growth of epiphytes on their leaves (Phillips, Eminson and Moss, 1978).

In deep lakes macrophytes are restricted to the littoral zone, but with enrichment marked changes occur in the species of plankton, and in their number. The plankton types associated with oligotrophic and eutrophic lakes are listed in Table 3.5. Desmids are particularly important in lakes low in nutrients, while blue-green algae often dominate lakes with high nutrient concentrations. The flora of diatoms also changes. Oligotrophic lakes have a diatom flora which is often dominated by species of *Cyclotella* and *Tabellaria*, while eutrophic lakes have *Asterionella*, *Fragillaria crotonensis*, *Stephanodiscus astraea* and *Melosira granulata* as dominants.

The diatoms are particularly interesting because they have siliceous frustules, which fall to the lake floor on the death of the cell. By examining a time series of sections through a sediment core it is possible to describe the changes which have occurred in the waters of a lake. Moss (1972) has examined a sediment core from Gull Lake, Michigan. When the earlier sediments were laid down, *Cyclotella michiganiana* and *Stephanodiscus niagarae* were dominant. From 20 cm in the sediment upwards they were joined by *Cyclotella comta* and *Melosira italica*. At 15 cm in the sediment a marked change occurred, with those species already present becoming much more abundant and *Asterionella formosa* appearing. At 10 cm *Fragillaria crotonensis* appeared while the abundance of the other species continued to increase. The total diatom density near the surface of the sediment was fifty times greater than in the lower layers. The sediment record thus reflected the increase in nutrient supply to the lake both in terms of a change in diatom species present and in an increase in diatom abundance.

Fig. 3.10. Photographs of Alderfen Broad taken in (a) 1930 and (b) 1975 showing loss of aquatic macrophytes. (Photo (a) from *Trans. Norf. and Norwich Nat. Soc.* **13**, photo (b) M. J. Hardy).

Table 3.5. Characteristic algal associations of oligotrophic and eutrophic lakes.

	Algal group	*Examples*
Oligotrophic lakes	Desmid plankton	*Staurodesmus*, *Staurastrum*
	Chrysophycean plankton	*Dinobryon*
	Diatom plankton	*Cyclotella*, *Tabellaria*
	Dinoflagellate plankton	*Peridinium*, *Ceratium*
	Chlorococcal plankton	*Oocystis*
Eutrophic lakes	Diatom plankton	*Asterionella*, *Fragillaria crotonensis*, *Stephanodiscus astraea*, *Melosira granulata*
	Dinoflagellate plankton	*Peridinium bipes*, *Ceratium*, *Glenodinium*
	Chlorococcal plankton	*Pediastrum*, *Scenedesmus*
	Myxophycean plankton	*Anacystis*, *Aphanizomenon*, *Anabaena*

The concentration of chlorophyll-*a* in the water is often taken as an index of the biomass of algae present. For oligotrophic lakes the mean summer concentration of chlorophyll-*a* in the epilimneon is in the range 0.3–2.5 mg m^{-3}, whereas for eutrophic lakes the range is 5–140 mg m^{-3} (Vollenweider, 1968). Enrichment also affects the rate of primary production. The mean daily rates of primary production are 30–100 mg C m^{-2} d^{-1} for oligotrophic lakes and 300–3,000 mg C m^{-2} d^{-1} for eutrophic lakes (Rohde, 1969), being equivalent to annual rates of 7–25 g C m^{-2} and 75–700 g C m^{-2} respectively.

With an increase in primary productivity we could reasonably expect an increase in the zooplankton. However, different types of algae are utilized to different extents by zooplankton. Compared with green algae and diatoms, blue-greens are assimilated with low efficiency (Schindler, 1971). The bottom dwelling cladoceran *Chydorus sphaericus* can eat blue-greens and its population often increases markedly during phytoplankton blooms (Brooks, 1969). Changes in the zooplankton community with enrichment may also be influenced by changes in the fish community, resulting in a shift in the pressure of predation.

The general changes in the fish fauna with eutrophication have been summarized by Hartmann (1977), using information from fifty-one European lakes. From the changes in the yield of fish Hartmann recognized four stages in eutrophication (Fig. 3.11). Coregonid fish

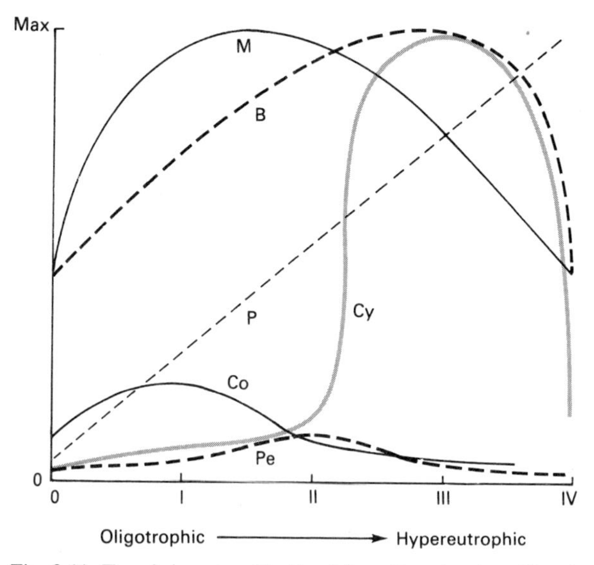

Fig. 3.11. Trends in eutrophicating lakes. B = density of benthos, Co = yield of whitefish (coregonids), Cy = yield of cyprinid fish, M = submerged macrophytes, P = phosphorus content of water, density of plankton and turbidity, Pe = yield of percid fish. O–IV = stages of eutrophication (from Hartmann, 1977).

dominate in oligotrophic lakes and their yield increases in the early stages of enrichment and then declines. There is some increase in the percids in the intermediate stages of eutrophication but they then decline. The yield of cyprinid fish increases sharply at intermediate stages of eutrophication and falls off sharply in heavily eutrophicated waters.

The reduction in diversity which occurs in the benthic fauna with enrichment has already been referred to but there may also be changes in the seasonal pattern of occurrence, which may have consequences for organisms higher in the food chain. The eutrophic, but unpolluted, Upton Broad mentioned earlier was dominated by larvae of the midge *Tanytarsus holochlorus*, with lesser numbers of caddis larvae, mayfly nymphs and snails, whereas the culturally eutrophic Alderfen Broad was dominated by the tubificid worm *Potamothrix hammoniensis* and the midge larva *Chironomus plumosus* (Mason, 1977a). The benthic populations in Upton Broad were highly seasonal and there was little biomass present in the autumn and winter. Alderfen Broad was markedly less seasonal, with a substantial biomass of benthic animals in autumn and winter. This is illustrated in terms of community respiration in Fig. 3.12. The fish in Upton

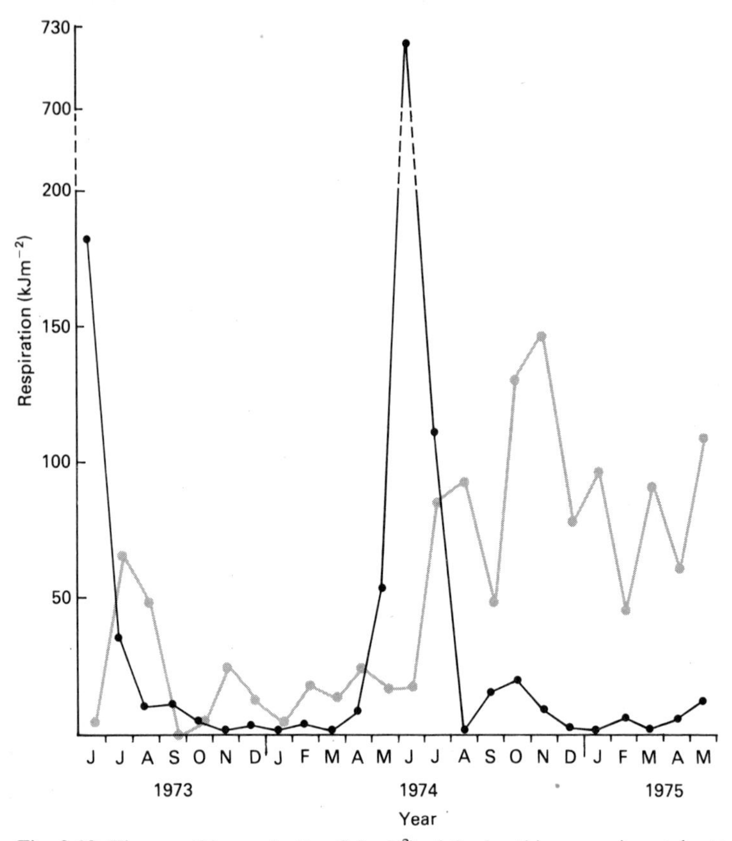

Fig. 3.12. The monthly respiration (kJ m^{-2}) of the benthic macro-invertebrate community at Alderfen Broad (○) and Upton Broad (●), eastern England, June 1973–May 1975 (from Mason, 1977).

Broad would probably have very little food from late summer to mid-spring and indeed the populations of fish were small and recruitment poor compared with high populations and good recruitment in Alderfen Broad.

Problems for man

The changes in nutrient levels and biology outlined above can create a number of problems directly affecting the activities of man (Table 3.4) and these can be summarized into three main areas (Wood, 1975)—problems associated with water purification, supply and consumption; problems associated with aesthetic and recreational activities and problems associated with the management of watercourses and lakes.

The increase in phytoplankton with eutrophication frequently causes severe problems in water purification. Water is treated either by double slow sand filtration, preceded by passage through rapid sand filters or microstrainers, or by coagulation and sedimentation, followed by rapid sand filtration. In the first method large algae are removed at the primary filter, which may become rapidly blocked at high algal densities, whereas large quantities of small algae can overload the slow sand filters. The coagulation and sedimentation in the second method of treatment are more efficient at removing large numbers of small algae, so that the residual large algae may block the rapid sand filters (Collingwood, 1977). The blocking of filters seriously reduces the throughput of water at the treatment works and occasionally a reservoir has to be taken out of service temporarily to clean the filters. Where the water supply for potable and industrial use in an area comes mainly from one large source the blockage of filters in the treatment works can be potentially very serious.

The smallest algal cells often pass through the filters and enter the supply to the consumer. The cells may decompose in the distribution pipes, allowing the growth of bacteria, fungi and some invertebrates (Collingwood, 1977). The resulting water can have an unpleasant taste and odour.

A water supply with a high nitrate level presents a potential health risk. In particular infants under six months of age may develop *methaemoglobinaemia* by drinking bottle-fed milk which is high in nitrates. Babies have gastric juices with a very low pH, which favours the reduction of nitrate ions to nitrite. Nitrite ions readily pass into the bloodstream, where they oxidize the ferrous ions in the haemoglobin molecules, reducing the oxygen carrying capacity of the blood (Beatson, 1978). Above 25 per cent methaemoglobin there is a blueing of the skin (cyanosis) and associated symptoms. Death occurs at levels of between 60 per cent and 85 per cent methaemoglobin. There have been some 2 000 cases notified worldwide, many of them fatal. The United Kingdom has had ten cases.

The European Health Standards for drinking water recommend that nitrate concentrations should not exceed 50 mg $NO_3\ l^{-1}$. Water remains acceptable at nitrate levels of 50–100 mg $NO_3\ l^{-1}$ (though six of the ten British cases occurred with water having nitrate levels less than 100 mg $NO_3\ l^{-1}$) but is unacceptable at levels above 100 mg $NO_3\ l^{-1}$. Standards in the United States are more stringent, with nitrate levels higher than 45 mg NO_3 l^{-1} unacceptable. Water derived from river intakes in lowland England frequently has nitrate levels exceeding the maximum recommended level (100 mg NO_3 l^{-1}) and has to be mixed with water low in nitrates before it enters the public supply. Alternatively, bottled water, low in nitrates, is supplied to mothers who are bottle-feeding young infants.

High levels of nitrates (greater than 100 mg NO_3 l^{-1}) in water may result in the formation of nitrosamines in the acidic human stomach. Nitrosamines are probably carcinogenic, but these effects have yet to be substantiated.

The aesthetic and recreational value of eutrophic waters is often reduced, though the detraction in amenity is a subjective assessment and many users are unaware of or unresponsive to the deterioration. There may be interference with fishing, sailing and swimming due to the production of surface scums during algal blooms and to the growth of algae and macrophytes on shores. The smell resulting from decaying algae washed ashore is often highly offensive. The dense swarms of midges which emerge from eutrophic lakes can be a considerable nuisance to lakeside visitors and residents, while the insecticides used to control midges cause damage to other organisms, including fishes, in the lake.

Management problems, other than water treatment, revolve around excessive weed growth and fisheries. The biomass of aquatic macrophytes increases with nutrient input and plants spread over lakes, rivers and canals, making navigation difficult and hindering recreation. Large growths of macrophytes impede the flow of water and increase the risk of flooding during storms. Many waterways have to be manually cleared of weeds annually during the summer.

It has already been recorded (p. 80) that increases in enrichment cause changes in the fish community. Salmonids and coregonids, which are high quality food fish are replaced by cyprinids, which are of low quality, though their biomass is usually higher. This change in dominant species is due to de-oxygenation of the hypolimnetic water. Low oxygen conditions may develop in eutrophic waters when algae and macrophytes die and decompose and this will also affect the numbers and species of fishes present. An increase in pH, associated with increased growth of plants, may also kill fish.

High densities of algae may produce toxins which are lethal to animals. An example is *Prymnesium parvum*, which grows well in nutrient-rich, slightly brackish waters and produces a toxin which is very potent to fish. It has caused problems in commercial fish-ponds in Israel and has severely damaged a first-class recreational fishery at

Hickling Broad, eastern England (Holdway *et al*, 1978). Fish kills have occurred in Hickling Broad since 1969 and the large populations of *Prymnesium* appear to be related to the increase in total phosphorus in the water. Populations of cells of 10^4–10^5 ml^{-1} are necessary to produce sufficient toxins to kill fishes and high toxin activity occurs when phosphorus becomes limiting, because the formation of phospholipids in cell membranes is disrupted, allowing leakage of the toxin (Dafni *et al*, 1972).

Similar problems arise with the bacterium *Clostridium botulinum*, which grows in the sediment of shallow, eutrophic lakes and releases a toxin during periods of hot weather. Birds are especially susceptible to botulism in shallow waters and serious losses of commercially valuable wildfowl periodically occur, especially in the USA (Smith, 1976).

These wide ranging effects of increases of nutrient loadings on receiving water bodies show why eutrophication is such an important problem today.

Experiments within lakes

Much of the information concerning eutrophication has been gained by comparing observations made on polluted and unpolluted waters or by comparing data collected from one site at successive intervals of time. However, it is often difficult to determine the cause-effect relationships in such a complex situation where many factors may vary simultaneously. The opposite approach is to conduct laboratory experiments, where cause-effect relationships can be verified by direct test, but it is often dangerous to apply results obtained under precise laboratory conditions to field situations (Johnson and Vallentyne, 1971). The manipulation of waters in the field lies between these two approaches and has proved highly illuminating in elucidating aspects of the eutrophication process.

Part of a lake may be isolated and nutrients added, using conditions in the water outside the enclosure as the control. Alternatively, a whole lake may be fertilized and the changes occurring can be compared with observations made before enrichment or with conditions in a neighbouring and similar, but unfertilized, lake.

Lund tubes

Dr J. W. G. Lund of the Freshwater Biological Association in England has developed the technique of using large enclosures to carry out nutrient manipulation experiments in lakes. The enclosures (Fig. 3.13) are made of butyl rubber and have a diameter of 45.5 m and a depth of 15 m. The first tubes were placed in Blelham Tarn, in the English Lake District, where they held a volume of water of more than 18 000 m^3. The bottoms of the tubes sink into the mud, isolating the enclosed sediment from that outside, while at the surface there is a raised, inflatable rubber ring, preventing mixing of the enclosed surface water with that outside.

After the installation of the first tube the conditions were monitored for thirty months and the physical conditions and phytoplankton quality remained similar inside and outside (Lack and Lund, 1974). The chief difference was that the enclosed water was isolated from the influences of the shallow littoral of the lake and from lake inflows. This led to the eventual reversal of the water within the tube to an oligotrophic state from a moderately eutrophic state as phosphates became bound to the sediments and were not replaced by inputs from the catchment. There was a corresponding change in the phytoplankton (Lund, 1975).

Single additions of phosphate to the tubes in Blelham Tarn resulted in an increase in the phytoplankton, suggesting that algal growth in the tarn is normally limited by phosphorus. At some periods, silicon had to be added to stimulate the growth of diatoms (Lund, 1978).

Fertilization of the water in a tube in autumn with phosphate, silicate and a ferric complex of ethylinediaminetetracetic acid (EDTA, a chelating agent) brought forward the spring diatom increase characteristic of north temperate lakes, so that the maximum was reached in February, at a time when the increase in the lake body was only just beginning (Fig. 3.14). This suggested that lake flushing,

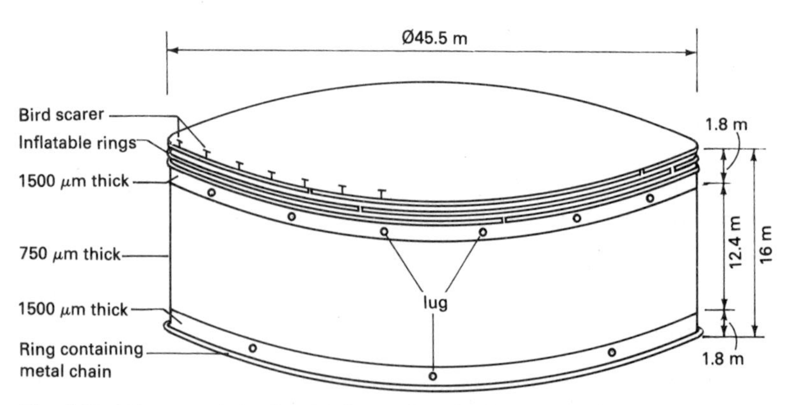

Fig. 3.13. Diagram of the Lund tube used in Blelham Tarn (from Lund, 1978).

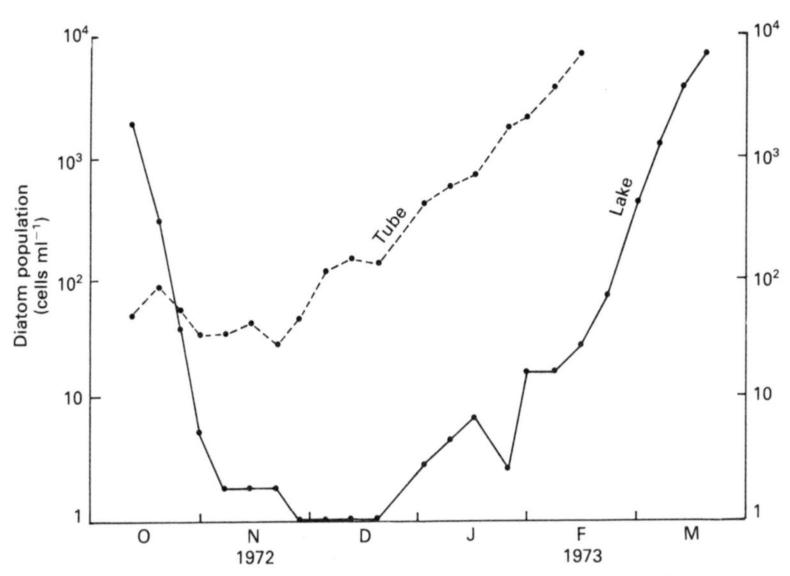

Fig. 3.14. Planktonic diatom populations in a Lund tube fertilized in September 1972 with silicate, phosphate and the ferric complex of EDTA compared with that in the lake water outside (from Lund, 1978).

caused by heavy winter rains, may have an important influence on the timing of the light-mediated, spring diatom increase, since nutrients added to the enclosures in autumn could not be flushed away (Lund, 1978).

The Lund tubes at the Freshwater Biological Association have also been used to investigate the ecology of bacteria, fungi and zooplankton with considerable success and similar tubes have since been installed by other workers at several sites (Lund, 1978).

The fertilization of Lake 227

Dr D. W. Schindler and his colleagues at the Freshwater Institute of the Fisheries Research Board of Canada have observed the effects of nutrient additions to whole lakes. Some forty-six small lakes, within an area containing several hundreds in Ontario, have been set aside for experimental research on eutrophication (Experimental Lakes Area) and Lake 227 was fertilized for four years beginning in 1969 (Schindler *et al*, 1973). Lake 227 has a surface area of 5 ha and a maximum depth of 10 m. Phosphorus and nitrogen were added weekly from 1969 to 1972, the annual addition amounting to

$0.48 \text{ gP m}^{-2} \text{ y}^{-1}$ and $6.29 \text{ gN m}^{-2} \text{ y}^{-1}$, these additions increasing the natural inputs by about five times.

The transparency of water, measured with a Secchi disc, became on average less in each year (Fig. 3.16) (Schindler *et al*, 1973). In 1969 the maximum biomass of phytoplankton was 16 160 mg m^{-3}, compared with 5000 mg m^{-1} in 1968, before fertilization. In 1970 the peak standing crop had almost doubled to 35 000 mg m^{-3}. Poor weather conditions in 1971 prevented a further increase in the maximum biomass, but the peak biomass in 1972 increased to 63 000 mg m^{-2}.

Before fertilization the phytoplankton of Lake 227 was dominated by Chrysophyceae. In 1969 Chlorophyta dominated in the summer. In 1970 blue-green algae, which had never occurred before fertilization, dominated in the late summer (Fig. 3.16). In 1971, blue-greens (especially *Oscillatoria*) again showed periods of dominance, although the situation was complicated by unusually cold weather in July and early August. In 1972 blue-green algae and Chlorophyta were again abundant, with the latter remaining dominant. The phytoplankton production in Lake 227 was some ten times higher than that occurring in natural lakes in the area (Schindler and Fee, 1973).

Other experiments in the Experimental Lakes Area have added elements singly. While both phosphorus and nitrogen were required to produce large standing crops of phytoplankton, the addition of

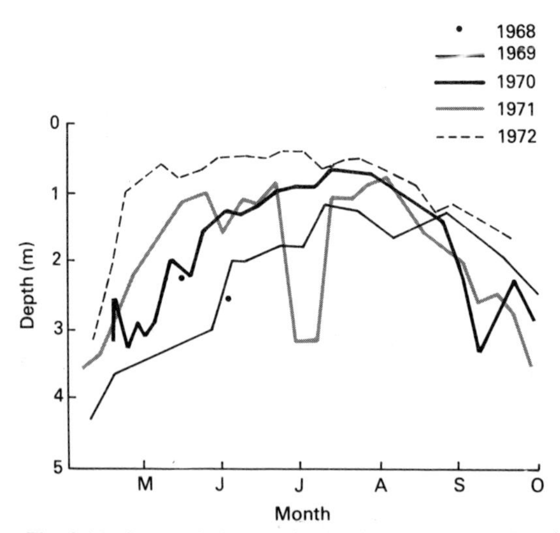

Fig. 3.15. Seasonal changes in the Secchi disc visibility in Lake 227, Ontario, Canada, 1968–1972 (from Schindler *et al*, 1973).

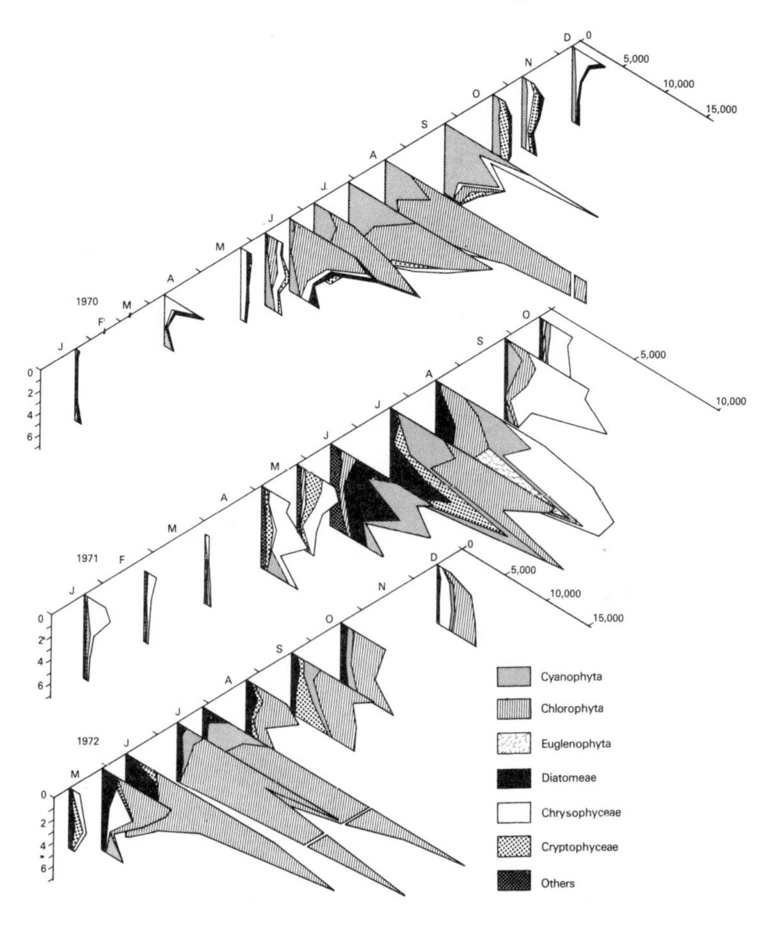

Fig. 3.16. The seasonal changes in abundance and depth distribution of major phytoplankton groups, Lake 227, 1970–1972. Depth in m (left-hand axis), abundance as cell volume $mm^3\ m^{-3}$ (right hand axis) (from Schindler *et al*, 1973).

phosphorus alone caused some increase, while nitrogen added alone never resulted in an increased standing crop (Schindler *et al*, 1973). As with the water in the Lund tubes it is the level of phosphorus which is controlling the standing crop of phytoplankton. Controlling the supply of phosphorus to lakes would therefore appear to be the key to controlling problems caused by eutrophication.

Changes in the fauna of lakes have been induced by artificial fertilization. Weglenska and Hillbricht-Ilkowska (1975) added nitrogen, phosphorus and potassium to four Polish lakes of different initial trophic status. Changes in the number and composition of the zooplankton occurred and these were most marked in the lake which was intially the least productive. Oligochaetes made up a greater proportion of the benthic biomass and chironomids a lesser proportion after fertilization, confirming general observations on the benthos of eutrophic lakes (e.g. Mason, 1977a, Carter, 1978). In general the Polish lakes became more uniform due to the eutrophication.

Modelling eutrophication

Many lakes are exhibiting eutrophication due to the activities of man and it is clearly impossible to study all of these in detail. Relationships which would allow the prediction of changes caused by rates of enrichment and requiring the measurement of as few parameters as possible would obviously be extremely valuable to the managers of freshwater ecosystems. Models of eutrophication vary widely in their complexity (e.g. Middlebrooks *et al*, 1974; Imboden and Gächter, 1978; Jørgensen *et al*, 1978; Scavia and Robertson, 1979). Vollenweider (1975) has emphasized that satisfactory models should meet three essential criteria. They should be general, realistic and precise. In practice one of these criteria usually has to be sacrificed in order to maximise the others.

Vollenweider (1968, 1969, 1975) has examined the input-output relationships of nutrients in lakes, with particular reference to phosphorus and nitrogen. Vollenweider related the loading rate (the annual input of a nutrient per unit lake area, L) to the mean depth of a lake ($\bar{z}$) and produced loading values of nitrogen and phosphorus which would be permissible for lakes of different depths, and loading values which would cause deterioration of the lake (Table 3.6). The ratio of nitrogen to phosphorus in the water can also provide useful information. If the N:P ratio exceeds 16:1, phosphorus is likely to be limiting to algal growth, if the ratio is less than 16:1, nitrogen may be limiting.

The simple relationship between loading and depth has been further developed by including such parameters as flushing rate (the water renewal time), the internal loading (such as from the sediments) and the length of the shoreline.

Vollenweider and Dillon (1974) plotted the relationship of L against $\bar{z}/T_w$, where T_w was the residence time of water in the lake. Residence time can be determined by dividing the lake volume (V) by the annual volume of the outflow (Q).

Table 3.6. Vollenweider's permissible loading levels for total nitrogen and total phosphorus (biochemically active), g m^{-2} y^{-1}.

	Permissible loading		*Dangerous loading*	
Mean Depth (m)	*N*	*P*	*N*	*P*
5	1.0	0.07	2.0	0.13
10	1.5	0.10	3.0	0.20
50	4.0	0.25	8.0	0.50
100	6.0	0.40	12.0	0.80
150	7.5	0.50	15.0	1.00
200	9.0	0.60	18.0	1.20

The retention time and flushing rate can be included in the relationship by plotting $L\ (1-R)/\rho$ against $\bar{z}$, where $1-R$ is that fraction of loading not retained in the lake and ρ is the flushing rate $(1-T_w)$.

Dillon and Rigler (1975) have used these relationships to predict the effect of development on lakes in Ontario and their study is worth examining in detail to show the value of relatively simple models. In Ontario there is a great demand for summer cottages and similar developments

The approach used by Dillon and Rigler is illustrated in Fig. 3.17. The *total phosphorus* supplied to a lake from natural sources (J_N) derives from the catchment (J_E) and direct precipitation (J_{PR}). An estimate for the export of phosphorus from a catchment can be made by examining the land use (e.g. the proportions of forestry, arable and pasture) and the geology (e.g. whether igneous or sedimentary).

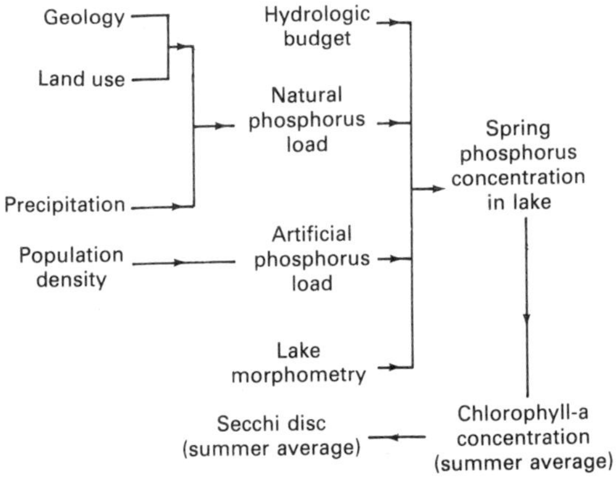

Fig. 3.17. Scheme of empirical models used to assess the effects of development on the trophic status of lakes (from Dillon and Rigler, 1975).

The export coefficient is multiplied by the area of the catchment and these are summed for all tributaries to the lake to obtain J_N. An export scheme for the Ontario area has been derived by Dillon and Kirchner (1975). J_{PR} can be determined from long-term average rainfall values and the average concentration of phosphorus in rainwater.

The total natural phosphorus supplied to a lake in a year is:

$$J_N = J_E + J_{PR} \text{ mg y}^{-1} \quad [3.1]$$

and the total natural loading (L_N) is J_N divided by the surface area of the lake (A_O) in mg m^{-2} y^{-1}.

The artificial loading (J_A) to the lake will depend on a number of factors, such as the number of cottages, caravans etc, their methods of sewage disposal, the number of inhabitants and the time they spend in the cottage each year. Dillon and Rigler assumed that the average North·American supplies 0.8 kgP to the environment each year. An estimate of the number and usage of cottages requires a survey.

The total supply of phosphorus to the lake (J_T) is:

$$J_T = J_N + J_A \text{ mg y}^{-1} \quad [3.2]$$

and the total loading (L_T) is:

$$L_T = J_T/A_O \text{ mg m}^{-2}\text{y}^{-1} \quad [3.3]$$

To predict the spring phosphorus concentration [P] in the lake, Dillion and Rigler used a model derived from Vollenweider (1969), which assumes that changes in the phosphorus concentration in the lake are equal to the supply added by unit volume minus the losses through sedimentation and via the outflow:

$$\frac{d[P]}{dt} = \frac{J_T}{V} - \sigma[P] - \rho[P] \quad [3.4]$$

where V is the lake volume, σ is the sedimentation rate (y^{-1}) and ρ is the flushing rate (y^{-1}), which is equal to Q/V. This equation can be solved in relation to time and for the steady state situation the solution is:

$$[P] = \frac{L_T}{\bar{z}(\sigma - \rho)} \quad [3.5]$$

after substituting $L_T(= J_T/A_O)$ and $\bar{z}$ $(= V/A_O)$.

If the loading, mean depth, sedimentation rate and flushing rate can be predicted, or easily measured, the phosphorus concentration in the lake can be determined. The sedimentation rate is difficult to determine directly, so is obtained indirectly through the retention coefficient, which is highly correlated with the areal loading and hence easily predicted.

Knowing the spring phosphorus concentration [*P*], the summer average chlorophyll-*a* concentration can be predicted, using a relationship derived by Dillon and Rigler (1974):

$$\log_{10}\,[\text{chl.}a] = 1.45\,\log_{10}\,[P] - 1.14\ \text{mg m}^{-3} \qquad [3.6]$$

and similar relationships have recently been extended by Schindler *et al* (1978). Dillon and Rigler have also shown a relationship between chlorophyll-*a* and transparency (the Secchi disc reading, Fig. 3.18).

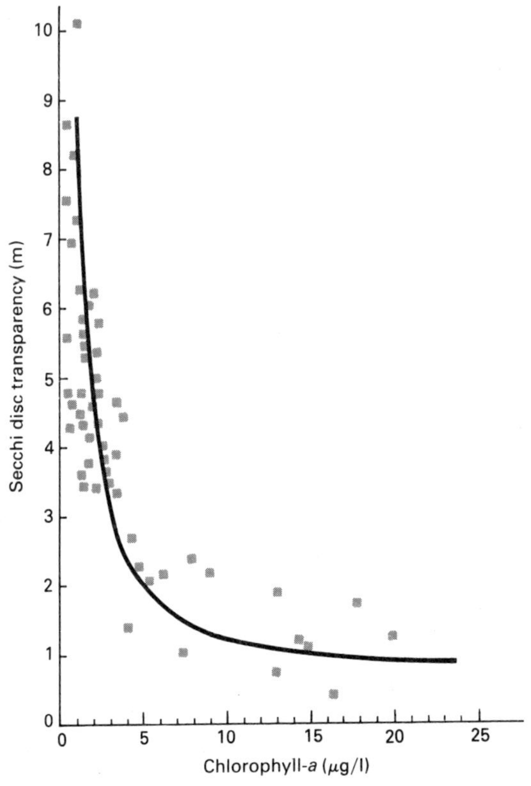

Fig. 3.18. The relationship between Secchi disc depth and chlorophyll-*a* concentration for a number of lakes in southern Ontario. The values for each lake are based on means collected over the period of stratification (June–September)(from Dillon and Rigler 1975).

Changes in the loading of phosphorus to a lake result in changes in the phosphorus concentration and hence water quality, but the response is gradual, rather than immediate, and follows an exponential relationship. The response time can be described by the half-life of the change in concentration ($t\frac{1}{2}$) which depends only on the rate coefficients and

$$t\tfrac{1}{2} = ln2(\sigma + \rho) \qquad [3.7]$$

Lakes with rapid flushing times will have short half-lives and hence response times, while lakes with slow flushing rates will respond only slowly to changes in loading.

The Dillon-Rigler model would most often be used by managers in the reverse direction. A minimum summer lake transparency or maximum chlorophyll-*a* concentration which is considered acceptable would be decided. The maximum permissible loading for phosphorus from artificial sources (L_A) would be determined using the model and this would be translated into the maximum allowable development in terms of cottages etc.

It would obviously not be feasible in terms of resources or time to scientifically evaluate every lake in Ontario where development was anticipated, so the model is valuable in that reasonable predictions of effects of development can be made with little or no field work. While some of the parameters used in the model apply only to Ontario, similar models could have wide application in the management of water resources.

Controlling eutrophication

We have seen that the productivity of freshwaters is most often limited by phosphorus. Eutrophication may therefore be most effectively controlled by limiting the loading of phosphorus. Thomas (1969) has discussed the rationale behind concentrating on phosphorus limitation. Phosphate is present in only trace amounts in oligotrophic lakes and the inflow streams are low in phosphate where they are not influenced by the activities of man. In contrast, the inflow streams may contain large quantities of nitrates. Nitrates are leached readily from agricultural lands, whereas phosphates tend to be tightly bound (p. 68). Some blue-green algae and bacteria can fix gaseous nitrogen (p. 64) so will not be limited by nitrogen provided other essential nutrients are available. It is also relatively easier and cheaper to remove phosphorus than nitrogen during the sewage treatment process. Nevertheless, where water is reused for potable supply it may be necessary to reduce the level of nitrogen for public health reasons and this has received considerable attention recently.

Nitrogen removal

Nitrogen is usually removed from wastewater by biological processes, involving nitrification and denitrification and, when preceded by secondary treatment, over 90 per cent removal of total nitrogen can be achieved. Nitrification involves the oxidation of ammonia to nitrate, with nitrite as an intermediate:

$$2NH_4^+ + 3O_2 \rightarrow 2NO_2^- + 2H_2O + 4H^+$$
$$2NO_2 + O_2 \rightarrow 2NO_3^-$$

The reactions are carried out by *Nitrosomonas* and *Nitrobacter* respectively. Denitrification involves the conversion of nitrate to nitrogen gas and a number of facultative heterotrophs use nitrate instead of oxygen as the final electron acceptor during the breakdown of organic matter. With methanol as the organic carbon source the reaction is:

$$6NO_3^- + 5CH_3OH \rightarrow 3N_2 + 5CO_2 + 7H_2O + 6OH^-$$

Because nitrified effluent contains little carbon, a carbon source is normally added. This is frequently methanol because it is almost completely oxidized, thus producing less sludge for disposal, and it is relatively inexpensive.

A number of tertiary treatment plants to remove nitrogen have been designed recently and they are reviewed in Wanielista and Eckenfelder (1978). A novel system is the biological fluidized bed in which wastewater is passed upward through a reaction vessel which is partially filled with a fine grained medium. The velocity is sufficient to fluidize the bed (i.e. impart motion to it) and a microbial community develops on the surface of the medium. Activated carbon and sand are typical media and the medium offers a large surface area per volume of reactor. As the particles are in fluid motion there is no contact between them, allowing the entire surface to be in contact with the wastewater. The fluidized bed combines the best features of the trickling filter and activated sludge plant and, under winter conditions, 99 per cent nitrification and denitrification can be achieved in 26 mins and 6.5 mins respectively (Jeris and Owens, 1978).

Phosphorus removal and the restoration of lakes

The discharge of nutrient-rich sewage and animal wastes into water-courses and lakes is often the most important source of enrichment problems, while detergents are often the most important source of phosphorus in waste waters. In the USA Dunst (1974) recorded that 40–70 per cent of phosphorus within sewage came from phosphate detergents (see also p. 66 for Britain) and the output from municipal sewage plants may contain up to 70 per cent of the total input of phosphorus to a lake. The elimination of phosphorus as an essential component of detergents could therefore remove about 50 per cent of the total phosphorus entering some lakes. Effective phosphate-free detergents have been developed and some states and cities in the USA have banned or restricted the phosphorus content of detergents.

Primary sewage treatment removes only about 5–15 per cent nutrients and secondary treatment only 30–50 per cent (Convery, 1970), so that a tertiary treatment process is necessary to remove the majority of phosphorus contained within sewage. Treatment may involve chemical removal, physical removal or biological removal, the former proving more practicable in most circumstances. Chemical precipitation with lime and iron salts, sometimes combined with filtration, is over 90 per cent efficient at removing phosphorus in the laboratory and in pilot plants, though the efficiency may be lower in larger, operational works. The operating costs of tertiary treatment may be equivalent to those of secondary treatment, thus doubling the overall costs of plants removing phosphate. It is therefore essential to know accurately the proportion of the total phosphorus loading due to sewage effluents and the efficiency of the proposed tertiary treatment plant before it is installed.

The reduction in nutrient loadings by altering land use practices may involve structural and land treatment measures to intercept nutrients and sediments before they reach water bodies, or may involve regulatory approaches, controlling land uses which directly or indirectly contribute to pollution (Dunst, 1974). The former measures involve the prevention of erosion, the efficient use of fertilizers and the development of methods of dealing with animal manures, especially from intensive livestock units. Some land use practices might be prohibited by law, while lakeside development could be controlled, possibly using schemes similar to those of Dillon and Rigler (1975) discussed above.

It is not always feasible to control nutrient inputs or it may be considered that the response time after nutrient limitation may be too long. For instance there may be an internal loading of phosphorus from the sediments. There is normally a net flow of phosphorus to the sediments but release occurs under conditions of low oxygen and this can stimulate algal growth. In shallow lakes the sediment is thought to contribute substantial amounts of phosphorus due to physical and biological activity, as well as to diffusion, and Swedish experience (Ryding and Forsberg, 1976) suggests that shallow, heavily polluted lakes are not significantly improved merely by reducing phosphorus income because of the high internal loading.

The modification of conditions within a lake can be achieved by dredging out the sediment, which both removes a source of nutrients and deepens the lake. If the deepening of a shallow lake is sufficient to allow the formation of a summer thermocline, there could be a substantial reduction in the quantities of nutrients reaching the euphotic zone during the main growing season for phytoplankton.

The process of lake recovery may be speeded up by inactivating or precipitating nutrients. The addition of aluminium sulphate or liquid alum to ponds and small lakes has given varying degrees of success (Dunst, 1974). Browman and Harris (1973, in Dunst, 1974) examined in the laboratory the effect of aluminium on phosphorus in intact lake cores and reported a good removal of inorganic phosphorus in water-sediment systems, with no removal of dissolved organic phosphorus. The aluminium floc which formed on the sediment surface suppressed the release of phosphorus from the sediment. This method may therefore have considerable potential, after further development, in small water bodies.

Other techniques to prevent recycling of nutrients or to accelerate the outflow of nutrients have included sealing lake bottoms (with polythene sheeting), selectively discharging hypolimnetic water in water supply reservoirs, or diluting and/or flushing with water from an oligotrophic source. Biotic harvesting, i.e. the removal of nutrient rich macrophytes, algae or fish has also been attempted.

The experience at Lake Washington has shown how a large, deep lake can be restored by diverting the incoming nutrients from waste water. Lake Trummen, in southern Sweden, which was formerly oligotrophic, was polluted with waste water from 1936 to 1958, after which the water was diverted (Gelin, 1978). Lake Trummen is shallow, with a maximum depth of 2 m. The recovery of the lake proved unsatisfactory and it was found that a black sediment, laid down during the period of pollution, was releasing nutrients and causing extensive blooms of blue-green algae (Andersson *et al*, 1973). During 1970 and 1971 the sediment was removed with a suction dredger and

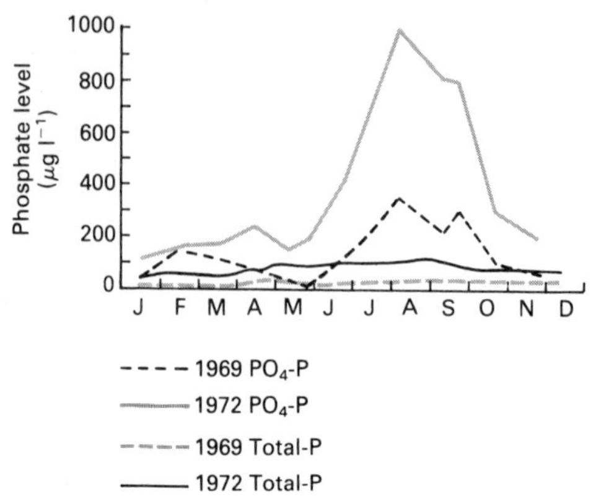

Fig. 3.19. Seasonal variation in phosphate and total phosphorus in Lake Trummen, Sweden, in 1969 and 1972 (from Andersson *et al*, 1973).

there was an immediate and striking improvement in the lake. The seasonal variations in phosphorus concentration were very much lower after dredging (Fig. 3.19), though there was still considerable algal growth. The transparency of the lake increased (Fig. 3.20) and the summer pH, previously often very high due to the intense algal production, was reduced. The annual phytoplankton production decreased from 345 gC m^{-2} lake surface in 1969 to 245 gC m^{-2} in 1972, with production occurring at a greater depth as the surface blue-green algae decreased in number. There were also marked differences in

Fig. 3.20. Seasonal variation in Secchi disc transparency in Lake Trummen, Sweden, in 1969 and 1972 (from Andersson *et al*, 1973).

the zooplankton. In 1972 the abundance of rotifers was only 41 per cent of the 1969 level, cladocerans were only 1.6 per cent and copepods only 64 per cent (Andersson *et al*, 1973). Species such as *Brachionus angularis*, *Keratella quadrata* and *Chydorus sphaericus*, considered to be indicators of eutrophy, showed marked declines in numbers. The total cost of restoring Lake Trummen, excluding research costs, was 580 000 dollars (Gelin, 1978).

The management and restoration of lakes has received a large amount of research in recent years (e.g. reviews by Jørgensen, 1980; Loehr et al, 1980).

Lakes can be restored successfully and Dunst (1974) listed almost 600 accounts of experience in lake restoration, the majority of which were ongoing projects.

Chapter 4

TOXIC POLLUTION

Introduction

The great variety of pollutants which affect the majority of water-courses receiving domestic, industrial or agricultural effluents has already been alluded to several times and these complex situations become especially apparent when considering toxic pollution. As an example, Klein (1962) described the rivers draining the urban conurbations of Manchester and Liverpool in north-west England as containing 'waste waters from tanneries, fellmongers and leather dressers; food processing; rubber proofing; gas-works; tar-distilling; electro-plating; iron pickling; coal washing; sand washing; quarrying; oil and grease processing and refining; the scouring of cotton and wool; the bleaching, finishing and macerizing of cotton and rayon; the dyeing of cotton, wool, jute and rayon; piggeries; slaughter-houses; calico-printing; and from the manufacture of batteries, paint, light alloys, concrete, rubber, plastics, rayon, dyes, chemicals, glue, gelatine, size, paper pulp and paper'. It is small wonder that the rivers in this area remain some of the most polluted in Britain!

Some of the industrial origins of toxic compounds are listed above, but agriculture and forestry also add many toxic pollutants to fresh-waters. These additions may be indirect, such as the run-off of insecticides and herbicides applied to the land, while waste pesticides and their empty containers are frequently carelessly dumped into ponds or streams with unfortunate effects. Toxic chemicals are also used in the direct control of particular members of the freshwater community. The most widely used are herbicides to control water plants considered to be interfering with man's use of freshwaters. Other organisms may be directly poisoned by the herbicide or may be indirectly affected by the change in community structure caused by the loss of plants. Insecticides are also applied directly to freshwaters, for instance to destroy the larvae of mosquitoes, the vectors of malaria. In tropical Africa, DDT is applied in large quantities to rivers to destroy the larvae of the blackfly, *Simulium damnosum*, the vector of the disease onchocerciasis, which in its most severe stage causes river blindness. Molluscicides (e.g. niclosamine) are widely used in the tropics to control the snail vectors of schistosomiasis (p. 20). In some countries of the world, piscicides are used to control fishes. This may involve the elimination of entire fish communities (e.g. with toxaphene), the elimination of selective groups of fishes (e.g. with antimycin), or the control of particular species,, such as larval sea lampreys (*Petromyzon marinus*) in the Great Lakes area of North America, using tri-fluoro methyl nitrophenol, TFM (Muirhead-Thomson, 1971).

Types of toxic pollutants

The major types of toxic pollutants can be listed as follows:

(a) metals, such as lead, nickel, cadmium, zinc, copper and mercury, arising from many industrial processes and some agricultural uses. The term 'heavy metal' is somewhat imprecise, but includes most metals with an atomic number greater than 20, but excludes alkali metals, alkaline earths, lanthanides and actinides.
(b) organic compounds, such as organo-chlorine pesticides, herbicides, polychlorinated biphenyls (PCBs), chlorinated aliphatic hydrocarbons, solvents, straight-chain surfactants, petroleum hydrocarbons, polynuclear aromatics, chlorinated dibenzodioxins, organometallic compounds, phenols, formaldehyde. They originate from a wide variety of industrial, agricultural and some domestic sources. The structure of some of these compounds is illustrated in Fig. 4.1.
(c) gases, such as chlorine and ammonia.
(d) anions, such as cyanides, fluorides, sulphides and sulphites.
(e) acids and alkalis.

It is generally considered that pesticides and metals head the list of environmental hazards at the moment (e.g. Korte, 1974), presenting a considerably greater potential risk than such pollutants as eutrophicating nutrients and organic sewage.

Environmental pollution by heavy metals became widely recognized with the Minamata disaster in Japan when, between 1953 and 1960, several thousand people suffered mercury poisoning from eating fish caught in Minamata Bay, which was receiving mercury

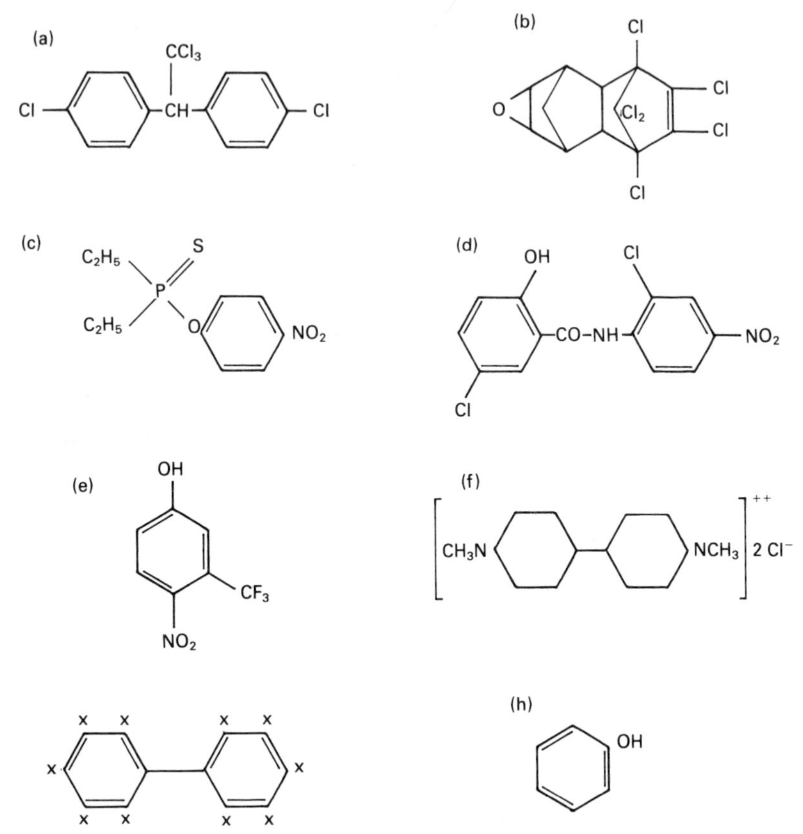

Fig. 4.1. The formulae of some organic compounds occurring as pollutants in aquatic ecosystems. (a) DDT; (b) dieldrin (both chlorinated hydrocarbon insecticides); (c) parathion (organo-phosphorus insecticide); (d) niclosamime (molluscide); (e) TFM (piscicide); (f) paraquat (herbicide); (g) polychlorinated biphenyl, chlorine atoms occur at any point marked X on the ring (in industrial effluents); (h) phenol.

released from a vinyl chloride plant (Smith and Smith, 1975). Minamata disease involved progressive loss of coordination, vision and hearing, together with intellectual deterioration. Similarly, the high level of cadmium in local foodstuffs in parts of Japan, attributable to irrigation water from the spoil heaps of a disused mine, caused Itai-Itai-Byo disease in 1955, mainly in women over forty. Itai-Itai-Byo disease is extremely painful and gives rise to bone deformities and fractures. The potential dangers of pesticides in the environment were brought to public notice in Rachel Carson's (1962) classic and controversial book, *Silent Spring*.

Table 4.1. The natural and man-induced rates of environmental contamination by some heavy metals, values in thousands of tons per year for 1967 (from Ketchum 1972).

Element	*Geological rate, G*	*Man-induced rate, M*	*M/G*
Tin	1.5	166	110
Lead	180	2330	13
Copper	375	4460	12
Manganese	440	1600	3.6
Mercury	3	7	2.3

Some potentially toxic compounds, such as heavy metals, are continually released into the aquatic environment from natural processes such as volcanic activity and weathering of rocks and a number (e.g. copper, zinc) are essential, in small amounts, to life. Industrial processes have greatly increased the mobilization of many metals (see Table 4.1), while the rate of manufacture of organic compounds is also steadily increasing (see Fig. 4.2). The world usage of pesticides is

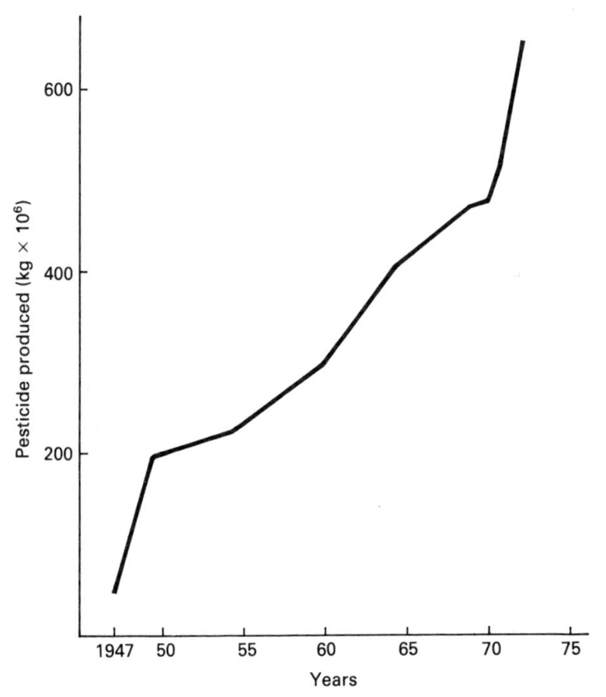

Fig. 4.2. Estimated amount of pesticide produced in the United States (adapted from Pimental and Goodman, 1978).

estimated at 1.8×10^9 kg, some 20 per cent of which is used in the United States (Fowler and Mahan, 1975) and the projected global usage in 1985 is estimated at 20 per cent above this figure.

The use of pesticides is reviewed in detail by McEwen and Stephenson (1979).

Toxicity

Alderdice (1967) has distinguished two general categories of toxic effect. *Acute* toxicity (a large dose of poison of short duration) is usually lethal, whereas *chronic* toxicity (a low dose of poison over a long time) may be either lethal or sublethal. Sprague (1969) has given a number of useful definitions for words in regular use in the study of toxic effects:

acute	coming speedily to a crisis.
chronic	continuing for a long time, lingering.
lethal	causing death, or sufficient to cause it, by direct action.
sublethal	below the level which directly causes death.
cumulative	brought about, or increased in strength, by successive additions.

There are also a number of terms which are used to express quantitatively the results of toxicity studies. These are, from American Public Health Association *et al* (1976):

Lethal concentration (LC) where death is the criterion of toxicity. The results are expressed with a number (LC_{50}, LC_{70}), which indicates the percentage of animals killed at a particular concentration. The time of exposure is also important in studies of toxicity so that this must also be stated, e.g. the 48-hour LC_{50} is the concentration of a toxic material which kills 50 per cent of the test organism in 48 hours.

Effective concentration (EC) is the term used when an effect other than death is being studied, for instance respiratory stress, developmental abnormalities or behavioural changes. The results are expressed in a similar way to lethal concentration (e.g. 48-hour EC_{50}).

Incipient lethal level is the concentration at which acute toxicity ceases, usually taken as the concentration at which 50 per cent of the population of test organisms can live for an indefinite time.

Safe concentration is the maximum concentration of a toxic substance that has no observable effect on a species after long-term exposure over one or more generations.

Maximum allowable toxicant concentration is the concentration of a toxic waste which may be present in a receiving water without causing harm to its productivity and its uses.

Acute toxicity

Examples of acute toxicity curves for fishes are illustrated in Figs. 4.3 and 4.4. Note that both axes are on a logarithmic scale. Figure 4.3 illustrates a curvilinear relationship, which has been observed for many toxic pollutants. The incipient LC_{50} can be obtained approximately as the asymptote and is about 25 mg l^{-1} for ammonia. Figure 4.4 shows a linear relationship with metals and there is an

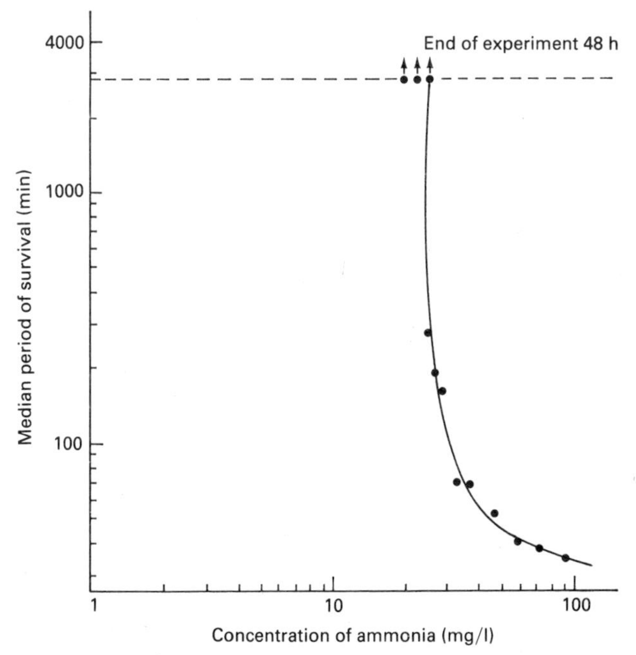

Fig. 4.3. A toxicity curve for trout in solutions of ammonia (NH_4Cl as mg l^{-1} N) at various concentrations. (from Herbert, 1961).

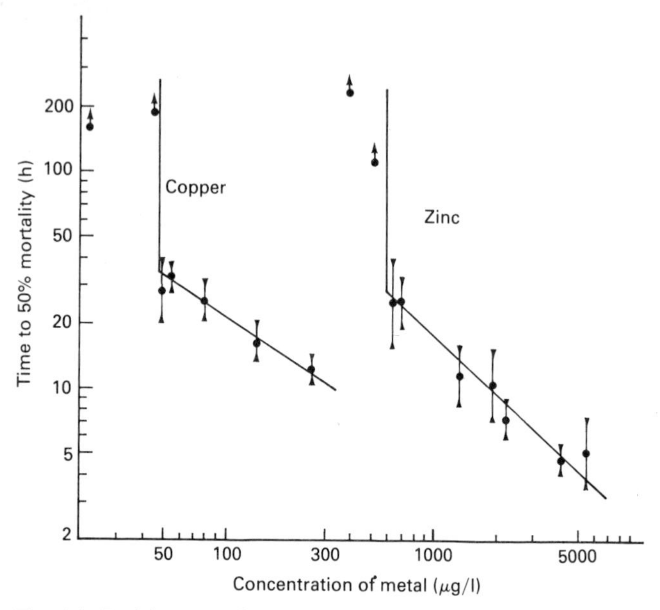

Fig. 4.4. Toxicity curves for salmon exposed to various concentrations of copper and zinc (from Sprague, 1964).

abrupt asymptote, the incipient LC_{50} being estimated at a concentration of 50 $\mu g\ l^{-1}$ for copper and 600 $\mu g\ l^{-1}$ for zinc. The incipient LC_{50} can be obtained more precisely using log-probit methods (Sprague, 1969). The incipient LC_{50} is a useful value in that it enables the toxicities of different pollutants to be easily compared and it forms a basic measuring unit for predicting the joint toxicity of two or more pollutants, as well as for describing sublethal effects (Sprague, 1969).

Different species vary in their vulnerability to specific pollutants. Solbé and Cooper (1976) compared the median lethal concentrations in hard water of three metals to the stone loach (*Noemacheilus barbatulus*) and rainbow trout (*Salmo gairdneri*) and the results are given in Table 4.2. Stone loach were more sensitive to zinc than rainbow trout, but they were more tolerant of cadmium. Both species showed similar sensitivities to copper. Muirhead-Thompson (1978a) found that the amphipod *Gammarus pulex* was 5000 times more tolerant than the mayfly nymph *Baetis rhodani* to the pesticide temephos. The sensitivity of individuals of a particular species to a pollutant may be influenced by internal factors, such as sex, age or size. For instance, the females of the crayfishes *Procambarus clarki* and *Faxonella*

Table 4.2. The median lethal concentrations ($mg l^{-1}$) of three metals to stone loach and rainbow trout (from Solbé and Cooper, 1976). Exposure time in days is in brackets.

	Copper	*Cadmium*	*Zinc*
Stone loach	0.26 (63)	2.0 (54)	2.5 (5)
Rainbow trout	0.28 (119)	0.017 (5.5)	4.6 (5)

clypeata were much more tolerant of mercury than the males (Heit and Fingerman, 1977). Females of both species survived throughout a 30-day exposure to 10^{-6} M mercuric chloride, whereas males suffered a 50 per cent mortality after only 3 days. Larger crayfish were better able to withstand high concentrations of mercury than small crayfish.

These differences in the tolerance of poisons between individuals of a species make it dangerous to extrapolate from simple laboratory toxicity tests on a standardized organism to the field situation. The test organism may appear tolerant, but a particular stage in its life cycle may be especially sensitive and this is the crucial stage in relation to the success of a population exposed to an environmental pollutant. The developmental or larval stages of an animal are generally more sensitive to toxic pollutants than adult stages. Blaylock and Frank (1979) found that the developing eggs of carp (*Cyprinus carpio*) were more sensitive to nickel toxicity than the newly hatched larvae, though they point out that, if the release of nickel was long term and chronic, the larval stage would be more crucial because the exposure time of the egg is curtailed by hatching. Another study of seven species of fish (not including carp) found that the eggs were more resistant than the larvae to four metals (Sauter *et al*, 1976) and similar results were obtained by McKim *et al*. (1978), who examined the toxicity of copper to the embryos and larvae of eight species of fishes. Biró (1979) looked at the acute effects of the herbicide 2–4D on the developmental stages of the bleak (*Alburnus alburnus*) and again found that larvae were more sensitive than the embryos, which at higher concentrations of herbicide either ceased developing or showed developmental abnormalities. Danil'chenko (1977) suggests that the greater resistance shown by early developmental stages is because the cytoplasm contains all the necessary resources for the morula and blastula stages, so that cleavage can occur under adverse conditions.

Mixtures of poisons

Effluents are often complex mixtures of poisons. If two or more poisons are present in an effluent they may exert a combined effect on an organism which is *additive*. Alternatively, they may interfere with one another (*antagonism*) or their overall effect on an organism may be greater than when acting alone (*synergism*). A generalized scheme describing the combined effects of two pollutants is given in Fig. 4.5. In this scheme the concentration of one unit of pollutant A produces the response in the absence of B and one unit of B does the same in the absence of A. If, on combining the two pollutants, the response falls within the square, joint action is occurring, with the pollutants aiding one another and this joint action can be broken down into three special cases. If the response is produced by combinations represented by points on the diagonal (e.g. 0.5A + 0.5B), the effects are additive. If the response is produced by combinations falling in the lower triangle of the box (e.g. 0.5A + 0.2B) the effect is more than additive (synergistic), while if the response falls in the upper triangle (e.g. 0.8A + 0.7B), the effect is less than additive though the pollutants are still working together in joint action (Sprague, 1970). Antagonism occurs in the region outside the box, e.g. when more than one unit of A is required to produce the effect in the presence of B. Antagonistic effects make the comparison between laboratory studies and field conditions difficult because animals in the field which

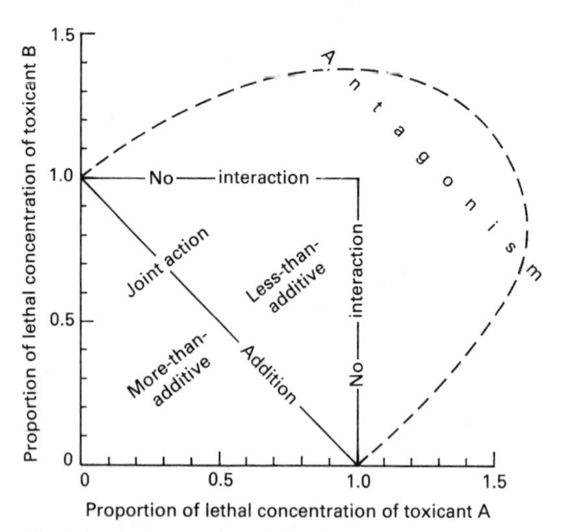

Fig. 4.5. The terms used to describe the combined effects of two pollutants (from Sprague, 1970, after Gaddum, 1948).

should theoretically be dead very often aren't. Anderson and D'Apollonia (1978) provide a review of multiple toxicity.

An example of an additive interaction is the combined toxicity of zinc and cadmium to fish, though their toxicity is synergistic to the alga *Hormidium rivulare* (Say and Whitton, 1977). Calcium is antagonistic to lead, zinc and aluminium. Copper is more than additive with chlorine, zinc, cadmium and mercury, whilst it decreases the toxicity of cyanide. Calamari and Marchetti (1973) have examined the toxicity of mixtures of metals and surfactants to rainbow trout. Figure 4.6 illustrates the toxicity of copper and a surfactant (sodium laurylbenzenesulphonate, LAS) acting separately and in combination and the toxicity can be seen to be markedly more than additive. It is generally considered that surfactants reduce the surface tension on the gill membranes, thus increasing the permeability of the gill to the surfactant and other poisons, but Calamari and Marchetti argue that the physical effects are less important than the chemical effects of the surfactant.

Sublethal effects

Poisons are frequently present in freshwaters at concentrations too low to cause rapid death directly, but they may impair the functioning

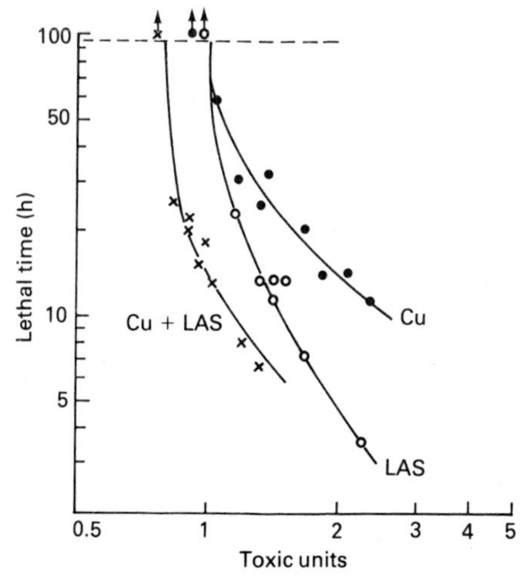

Fig. 4.6. Toxicity curves to rainbow trout of copper (Cu) alone, surfactant (LAS) alone and mixtures in various ratios of the two poisons (Cu-LAS). Lethal time to 50 per cent of animals in hours (h) and concentration in toxic units (from Calamari and Marchetti, 1973).

of organisms. These sublethal effects may be observed at the biochemical, physiological, behavioural or life-cycle level. Many small changes have been related to pollution but, as argued by Mount and Stephan (1967) and Sprague (1971) it is essential to show that these changes have ecological meaning, that they reduce the fitness of an organism in its environment and are not merely within the organism's range of adaptation. Sprague (1971) considers the biochemical effects of pollution as basic and these can then be related to the efficiency of tissues and organs, which can in turn be examined in relation to the performance of the animal and whether this has any adverse effect on the natural populations of the organism.

Christensen *et al* (1977) have argued that the early detection of specific molecular abnormalities in the tissues of organisms could provide an indication of exposure to pollutants long before any gross signs become apparent and such biochemical indices may be valuable in signalling the development of sublethal abnormalities which could reduce the fitness of a population. Christensen *et al* examined the effects of salts of methyl mercury, cadmium and lead on six biochemical parameters in the brook trout (*Salvelinus fontinalis*), with exposure periods from six to eight weeks. They observed that lead caused statistically significant increases in levels of plasma sodium and chloride, with decreases in haemoglobin and glutamic dehydrogenase activity. Cadmium caused increases in plasma chloride and lactic dehydrogenase activity, while decreasing plasma glucose levels, whereas methyl mercury resulted in increases in haemoglobin and plasma sodium and chloride.

Lead is known, amongst other things, to inhibit enzyme activity and acts at a large number of biochemical sites. Sastry and Gupta (1978) investigated the effect of sublethal concentrations of lead nitrate on digestive enzymes in the fish *Channa punctatus*. They found a decrease in the activity of alkaline phosphatase after 15 days exposure to lead, but there was no significant change after 30 days. Acid phosphatase showed elevated activity after 15 and 30 days and this was accompanied by cellular damage. Three carbohydrases showed elevation after 15 days, followed by inhibition after 30 days, whereas trypsin and pepsin were elevated. It was found that the activities of enzymes were altered in different ways in the liver and digestive system. In the same species of fish, Verma *et al* (1979) recorded a 65 per cent inhibition of oligomycin-insensitive Mg^{++} ATPase in the brain and gills caused by synthetic detergents.

The biochemistry of plants can also be adversely affected by toxic pollutants. Statton and Corke (1979) report that nitrogenase activity, measured in terms of acetylene production, was completely inhibited at a level of 20 mg l^{-1} cadmium in the blue-green alga *Anabaena inaequalis*, while 1 mg l^{-1} cadmium significantly inhibited photosynthesis, 20 mg l^{-1} resulting in complete inhibition of carbon dioxide fixation. Copper similarly affects photosynthesis in blue-greens, a concentration of 0.1 mg Cu l^{-1} reducing photosynthesis by 80 per cent in *Spirulina platensis* (Källquist and Meadows, 1978).

Pollutants act at the biochemical level at a number of sites but an organism may be able to adapt by normal homeostatic mechanisms so that enzyme inhibition may not reduce the overall fitness of an organism. Enzyme bio-assays remain, however, a useful technique in looking for sublethal effects of toxic pollution.

Figure 4.7 illustrates a generalized relationship between physiological impairment following increasing exposure to pollutants and the consequent disability of organisms. At low levels of pollution the organism is maintained in health by normal homeostatic mechanisms. As levels rise compensation occurs such that normal functioning is maintained without significant metabolic cost, but at higher levels still the organism becomes stressed and physiological breakdown occurs, the organism is unable to repair the damage and becomes disabled. Still higher loadings of pollutants result in physiological failure and death.

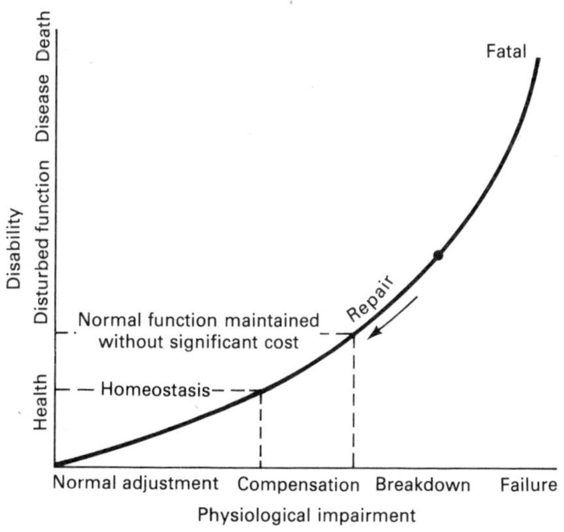

Fig. 4.7. The relationship between physiological impairment following increasing exposure to pollutants and the consequent disability of an organism (from Lloyd, 1972, after Hatch, 1962).

The effects of toxic pollutants on the respiration of fishes and invertebrates have received widespread attention and are reviewed respectively by Hughes (1976) and Wright (1978). By cannulating the blood system of fishes it is possible to measure the concentrations of oxygen, metabolites and pollutants and hence understand more fully the mode of action of toxic pollutants. Skidmore (1970), using cannulation techniques, found that zinc reduced the oxygen level of blood leaving the gills, but zinc injected into the blood system had no effect. Zinc, therefore, reduces the efficiency of oxygen transport across the gill membrane so that the fish die of hypoxia. Sellers *et al* (1975) have shown that zinc also reduces the pH of the blood, probably because metabolic products such as lactic acid are increased (Fig. 4.8).

Lloyd and Swift (1976) have examined the effects of pollutants on the water balance of fishes, using an indirect method whereby the urine flow rate was measured in rainbow trout fitted with a urinary catheter. The results illustrating the effects of various concentrations of ammonia on urine production are shown in Fig. 4.9. Urine flow rate increased with an increase in the external ammonia concentration and all fish died at the highest concentration of 20 mg l^{-1} N. The urine flow rate decreased in the lower ammonia concentration after one day, suggesting that acclimation was taking place, possibly by an increased rate of detoxification of ammonia, which might be depen-

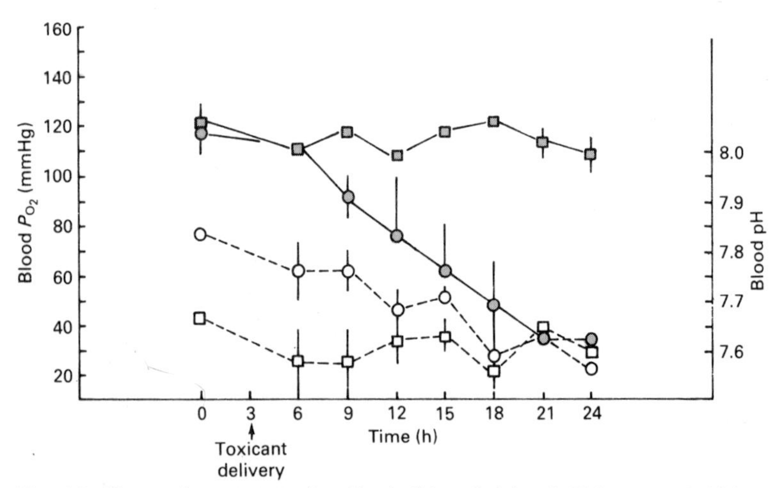

Fig. 4.8. Changes in oxygen tension, P_{O_2} (solid symbols) and pH (open symbols) in blood of rainbow trout treated with 1.43 mg l^{-1} zinc. Control values are represented as squares, experimental values as circles (after Sellers *et al*, 1975).

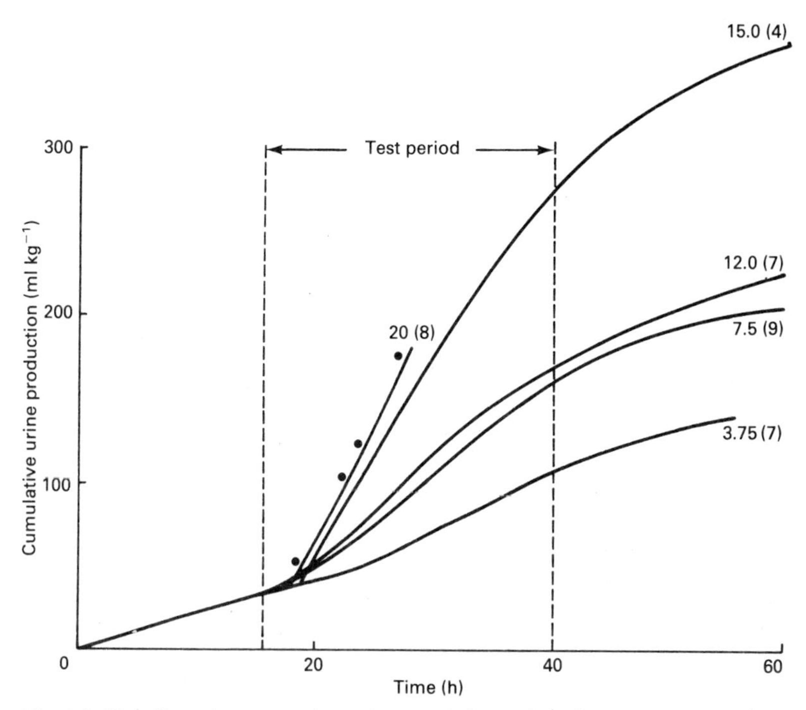

Fig. 4.9. The effect of concentrations of ammonia (as mg l^{-1} N) on urine production in rainbow trout. Ammonia concentrations are given against each curve, with the number of fish in brackets. The times for death at the highest concentration are shown as crosses. Temperature 10.5 °C, pH 8.1 (from Lloyd and Orr, 1969).

dent on an increased synthesis of glutamic acid, which is converted to glutamine in the presence of ammonia. The mechanism by which ammonia increases the water uptake by fish is unknown, though it could be related to increased activity, which may increase permeability. Smart (1978) has suggested that ammonia disturbs cellular metabolism, resulting in an increasing oxygen requirement, the hypermetabolic state accelerating the utilization of tissue energy reserves. High energy compounds may become depleted in the brain, as occurs in mammals, and the characteristic symptoms, including hyperexcitability and hyperventilation, have been observed in fish.

Growth can be examined as an integrated measure of the sublethal effects of toxic pollutants on the biochemistry and physiology of organisms but, with homeostatic mechanisms operating, the effects on growth may be minimal. Stoner and Livingston (1978) recorded no effect of bleached kraft mill effluent on growth, measured as length, in pinfish (*Lagodon rhomboides*), whereas Webb and Brett

(1972) recorded reduced growth in sockeye salmon (*Onchorhynchus nerka*) and McLeay and Brown (1974) reported stimulated growth in the coho salmon (*O. kisutch*). Examining the sublethal effects of zinc, Edwards and Brown (1966) found no difference in the growth rate between experimental and control rainbow trout and a similar absence of response was recorded by Farmer *et al* (1979) for Atlantic salmon (*Salmo salar*). However, both bluegills (*Lepomis macrochirus*) and fathead minnows (*Pimephales promelas*) exhibited retarded growth in zinc (Pickering, 1968, Brungs, 1969) and the pesticides endrin and malathion reduced growth in the flagfish (*Jordanella floridae*), according to Hermanutz (1978).

Although overall growth may not be affected by pollution, the composition of the body may be changed. Thus in the study by Stoner and Livingston (1978), referred to above, exposure of pinfish to 1 per cent bleached kraft mill effluent resulted in a reduced total lipid content and an increased protein content. The total protein content of the midge (*Chironomus tentans*) larva was reduced when exposed to cadmium (Rathore *et al*, 1979). The effects of sublethal levels of pollutants on growth is obviously not clear cut, but, as Sprague (1971) pointed out, it is easily measured and is an important criterion of success in natural populations and so should be recorded in laboratory studies of the effects of pollution.

Sublethal concentrations of toxic pollutants may affect the behaviour of organisms and the disruption of behaviour may reduce the fitness of a natural population. The swimming performance of fish may be reduced, caused by a reduction in oxygen uptake, as has been described for sublethal concentrations of pulpwood fibre (MacLeod and Smith, 1966). Subacute concentrations of DDT caused changes in the orientation behaviour, co-ordinated movements, feeding, aggression and comfort behaviour of the fish *Therapon jarbua* (Lingaraja *et al*, 1979). Chlorine and chloramine have been found to reduce the feeding rate of the rotifer *Brachionus plicatilis* (Capuzzo, 1979), while Bardach *et al* (1975) found that detergents damaged the taste buds of catfish *Ictalurus natalis* so that they were unable to find food. The precopulatory behaviour of the amphipod *Anisogammarus pugettensis* was impaired in the presence of bleached kraft pulp mill effluent, with rapid separation of paired animals in 40 per cent of full strength effluent (Davis, 1978). Animals may also avoid polluted water and this behaviour can be studied in the laboratory (Larrick *et al*, 1978). For example, the amphipod *Gammarus pulex* is very sensitive to pesticides and even low levels cause a marked increase in the downstream drift of this species (Muirhead-Thomson, 1978b).

Warner (1967) has reviewed the subject of sublethal pollution and behaviour.

By studying an organism under experimental conditions over its lifetime it is possible to find the weak link in its response to pollution, a technique developed by Mount and Stephan (1967). For example, at concentrations of copper greater than 0.03–0.07 of the 48 h LC_{50} and of zinc greater than 0.009 of the 48 h LC_{50}, the breeding success of fathead minnows is inhibited even though growth rates were normal (Mount, 1968; Brungs, 1969). Such long-term experiments are essential in order to discover any carcinogenic, teratogenic and mutagenic effects of pollutants.

Environmental factors affecting toxicity

Environmental factors may markedly modify the acute toxic effect of pollutants. Temperature is important because it not only influences the metabolic activity and behaviour of organisms, which may affect their exposure to a pollutant, but it may also alter the physical and chemical state of the pollutant. The effects of temperature on toxicity have been reviewed by Cairns *et al* (1975). In general, toxicity increases with temperature. Between 12 °C and 17 °C the toxicity of antimycin to goldfish (*Carassius auratus*) increased tenfold (Muirhead-Thomson, 1971), though Smith and Heath (1979), examining the effect of four pollutants on five species of fish, found that LC_{50}s did not differ by more than a factor of 3 over a 25 °C range. There are, however, many exceptions to the increase in toxicity with temperature. Brown *et al* (1967) observed that the time to death of rainbow trout exposed to phenol increased as temperature increased, but the LC_{50} decreased. Phenol causes paralysis and cardiovascular congestion, resulting in suffocation. The internal concentration of phenol is influenced by the relative rates of absorption and detoxification, both of which are directly proportional to temperature, but it is considered that temperature influences the rate of detoxification to a greater extent that the rate of absorption, at least at lower temperatures. Phenol therefore probably accumulates to higher levels at low temperatures, accounting for the greater toxicity in the cold.

The toxic effect of pollutants varies with the quality of water, pH and hardness being especially important. Hydrogen cyanide, for example, is especially toxic in the molecular form, so that any change in the pH which reduces the degree of dissociation will increase the toxicity of the solution without there being any change in the total concentration of cyanide. The toxicity of ammonia is also affected by pH.

An examination of the effect of hardness of water on toxicity is usually carried out after stabilizing the pH. Pollutants tend to be more toxic in soft waters (e.g. lead, copper, zinc). Figure 4.10 shows the relationship between the median survival time of rainbow trout and the concentration of zinc ions in waters of different total hardness (Lloyd, 1960). There was a linear relationship between survival and concentration of zinc in soft water, whereas the relationship was curvilinear at intermediate and high levels of hardness. Pollutants may be precipitated in hard water conditions (e.g. lead) and soluble complexes may also be formed. The toxicity of zinc decreases with an increase in inorganic suspended solids as the metal is absorbed by or adsorbed on to the suspended particles. At pH 7, Learner and Edwards (1963) found that copper sulphate was more toxic to the oligochaete worm *Nais* in soft water than in hard water, while in soft water it was much less toxic at pH 4 than at pH 7. Howarth and

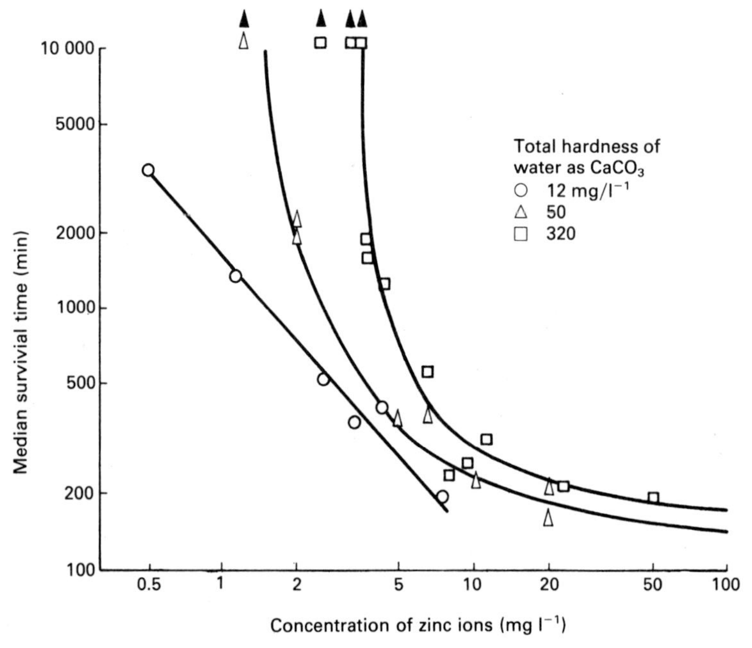

Fig. 4.10. The relationships between median survival time of rainbow trout and the concentration of zinc ions at three levels of hardness (after Lloyd, 1960).

Sprague (1978) determined the acute lethality of copper to rainbow trout in various concentrations of hardness and pH and some of the results are summarized in Fig. 4.11, which relates the lethal concentrations of ionic copper (Cu^{++}), plus $CuOH^{+}$ and $Cu(OH)_2^{++}$, considered to be the toxic species, to any combination of water hardness and pH. High hardness decreased the toxicity at any pH. The lethal concentration ranged from 0.09 $\mu g\ l^{-1}$ of copper in the softest and most alkaline water to 232 $\mu g\ l^{-1}$ in the hardest and most acid. Copper appeared to be especially toxic at pH 9 and Howarth and Sprague, following Lloyd and Herbert (1962) suggested that hydroxyl copper complexes accumulate between the gill filaments. Excretion of carbon dioxide at the gill surface lowers the pH in the immediate area, which ionizes hydroxides and liberates large quantities of cupric ions, which are taken up through the gill.

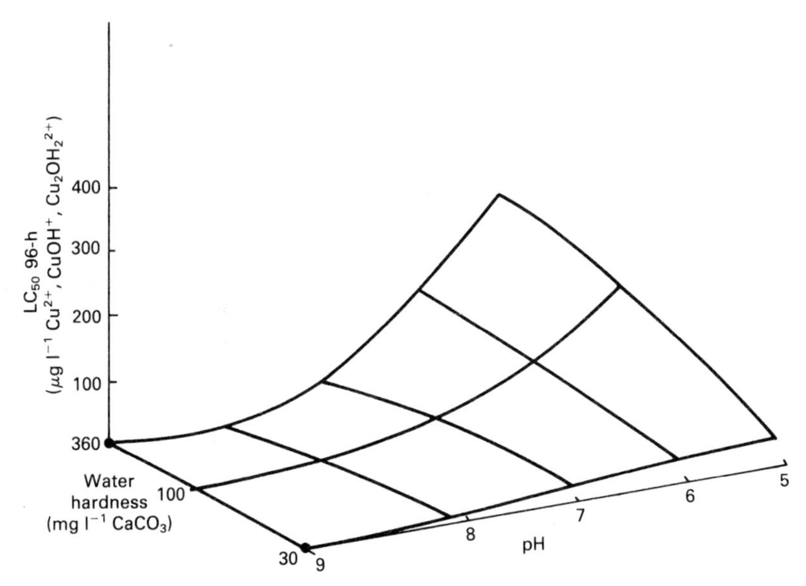

Fig. 4.11. The lethal concentrations of ionic copper to 10 g rainbow trout at various combinations of water hardness and pH (after Howarth and Sprague, 1978).

A number of poisons become more toxic at low oxygen concentrations because an increase in respiratory rate occurs, increasing the amount of poison being exposed to the animal. Lloyd (1961) observed that the toxicity of several poisons to rainbow trout increased in direct proportion to the decrease in oxygen concentration of the water.

Transformations

Many of the compounds released into watercourses are subject to transformations within the environment and this may render them more toxic. Mercury can be taken as an example and some of its transformations are illustrated in Fig. 4.12. Inorganic mercury is converted to methyl and dimethyl mercury in aquatic environments due to the activities of bacteria and fungi:

$$Hg^{++} \rightarrow CH_3Hg^+$$

The methylation process may occur under anaerobic (e.g. by *Clostridium*) or aerobic (e.g. by *Neurospora*, *Pseudomonas*) conditions. Methyl mercury is exceptionally toxic to many animals. Several non-biological transformations of mercury also occur, depending on the environmental conditions.

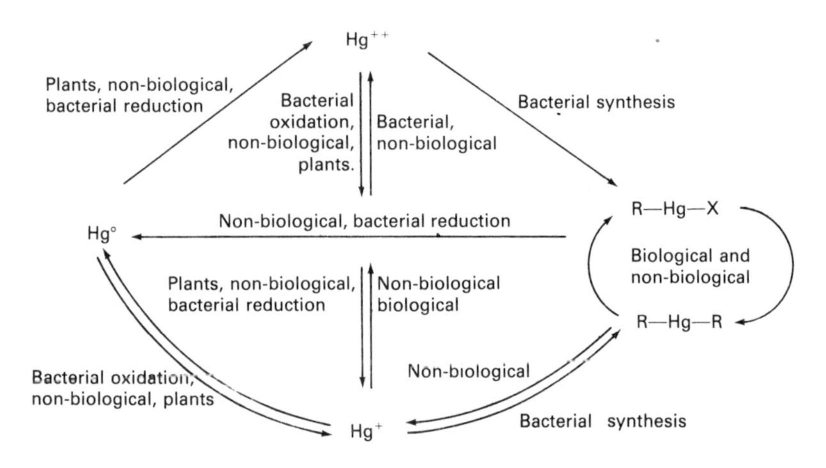

Fig. 4.12. Transformations of mercury (from Higgins and Burns, 1975).

Pesticides also undergo transformations in the environment, but these tend to be minor. For instance, aldrin is converted to dieldrin and DDT to DDE, but these products are still highly toxic. Chlorinated hydrocarbons (including PCBs) are very persistent in the environment and remain one of the great problems in environmental pollution. Recent reviews of the role of micro-organisms in the transformation of industrial chemicals and heavy metals have been produced by Colwell and Sayler (1978) and Iverson and Brinckman (1978).

Tolerance

Populations can develop a tolerance to pollutants which enables them to survive in highly polluted environments. They may achieve this by functioning normally at high toxic loadings or by metabolizing and detoxifying pollutants. The mechanisms of tolerance to pollution are extremely complex, involving several metabolic systems and species have solved the problem of tolerance to a particular pollutant in different ways.

In a study of Welsh rivers polluted with heavy metals from old mine workings, McClean and Jones (1975) found the filamentous green alga *Hormidium rivulare* to be highly tolerant of metals, while *Scapania undulata* was a highly tolerant moss. The moss *Fontinalis antipyretica* was less tolerant. In a wide-ranging survey of forty-seven sites with different levels of zinc, Say *et al*(1977) found *Hormidium rivulare* abundant at all sites, ranging from low zinc concentrations to as high as 30.2 mg l^{-1} Zn. The closely related *H. fluitans* occurred at sites with zinc only up to a mean of 5.59 mg l^{-1} Zn. Say *et al* showed that *Hormidium* growing in streams high in zinc were more tolerant than populations taken from unpolluted waters and this adaptation was genetically determined.

Brown (1977, 1978) has examined populations of the isopod *Asellus meridianus* which were tolerant to copper and lead. Copper tolerant animals could accumulate copper from solution and from the food to levels which proved lethal to non-tolerant animals, while a lead-tolerant population accumulated concentrations up to 30 $\mu g\ g^{-1}$ Pb, non-tolerant populations dying before tissue levels had reached 15–20 $\mu g\ g^{-1}$ Pb. Tolerance to copper in *A. meridianus* appears to confer tolerance to lead. The trace metals are stored in the hepatopancreas of *Asellus* and copper and lead compete for sites, with lead being more readily bound. The regulation of metals within aquatic invertebrates is reviewed by Wright (1978).

A previous exposure of an organism to low levels of a pollutant may make it more tolerant later. Beattie and Pascoe (1978) showed that pretreatment of rainbow trout eggs with cadmium gave protection to hatching larvae when they were exposed to cadmium. Pascoe and Beattie (1979) pretreated rainbow trout at 0.001 and 0.01 mg l^{-1} Cd and the 48 h LC_{50} on later exposure to cadmium was 0.11 and 1.5 mg l^{-1} Cd respectively. Control trout, not pretreated with cadmium, had a 48 h LC_{50} of less than 0.1 mg l^{-1} Cd when exposed later. It was suggested that pretreatment with low doses of metal stimulated the synthesis of a metal-binding protein, which could subsequently bind large doses of the metal to produce an inactive complex.

Many bacteria quickly become tolerant to toxic pollutants and the resistance factor is carried in the plasmids. Plasmids may transfer from cell to cell by bacterial conjugation or by transduction, involving transfer of viruses, and the transferred plasmid replicates rapidly and is passed on to the progeny, so that resistance can spread very quickly. Resistant bacteria might enhance the transport of pollutants, especially metals, in the environment by methods such as solubilization, concentration and conversion to organometallic species and their respective elemental forms, this constituting a potential environmental danger (Iverson and Brinckman, 1978). However, resistant bacteria can also remove heavy metals from the environment and, with the application of modern genetic engineering techniques, bacterial strains could be produced which may play a valuable role in de-toxifying industrial wastes.

Accumulation

Many toxic pollutants are concentrated along a food chain and very high levels can be accumulated in organisms from very low concentrations in water. Brooks and Rumsby (1965), for example, have observed concentration factors in the oyster *Ostrea sinuata* of 1.15×10^5 for zinc, 3.18×10^5 for cadmium and 4.0×10^3 for lead. The ability to concentrate pollutants varies between taxa. Molluscs accumulate cadmium to greater levels than crustaceans, which in turn concentrate more than fish (Frazier, 1979). Brooks and Rumsby (1965) suggested several possible pathways for the concentration of metals in aquatic ecosystems:

1. particulate ingestion of suspended material from the surrounding water;
2. ingestion of pollutants via their preconcentration in food;
3. complexing of pollutants with appropriate organic molecules;
4. incorporation of pollutants via physiologically important systems;
5. uptake of pollutants by exchange.

The relative importance of the different routes of uptake will depend on the biology of the species. For example, the aquatic plant *Nitella flexilis* accumulates metals primarily from the water, whereas accumulation from the sediment is also important in *Glyceria fluitans* (Harding and Whitton, 1978) and *Elodea canadensis* (Mayes *et al*, 1977).

The rate of accumulation of pollutants will depend on factors both external and internal to the organism. The level of pollution in the water is important and many species carry higher loadings of pollutants when living in polluted waters, though this is by no means always the case. Cember *et al* (1978) found that the bio-concentration of mercury in the fish *Lepomis macrochirus* increased exponentially with water temperature. Water chemistry also affects uptake and Frazier (1979) found that the cadmium level in molluscan tissues was inversely related to environmental salinity. Body tissue composition may also be important and Sugiura *et al* (1978) observed that the concentration of lipophilic organochlorine compounds in different species of fishes was directly related to the fat content of the fish.

The accumulation of a poison is a function of both uptake and elimination and a generalized curve is shown in Fig. 4.13. Assuming uptake to be due solely to chemical diffusion, the process will continue until the internal level is equal to the level in the environment (point X in Fig. 4.13). Most pollutants can, however, be eliminated from the body and this is an active biochemical and physiological process which cannot be described in simple diffusion terms. Both uptake and elimination occur simultaneously and the plateau will depend on the balance between factors determining the two processes.

Moriarty (1975a,b) has used a compartmental model to examine the quantitative relationship between exposure and the levels of pollutants in different tissues, with particular reference to DDT and its residues. In the simplest state a unicellular organism

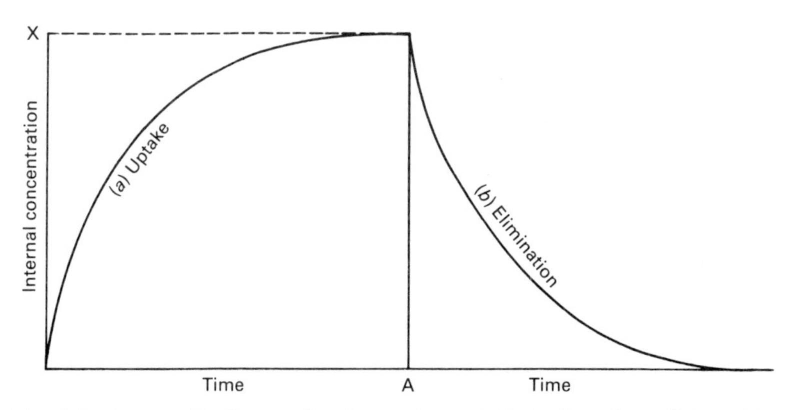

Fig. 4.13. A generalized curve for the uptake and elimination of a pollutant. The internal and external concentrations are equal at X, external concentrations are reduced to zero at time A (from Holdgate, 1979).

(compartment 1) contains a certain amount of residue (Q). When suspended in unpolluted water, the residue will be lost at a rate directly proportional to the amount present, assuming no metabolism, and the rate constant k_{01} is the amount excreted per amount present per unit time. The subscript indicates that the residue is passing from compartment 1 (the organism) to the compartment 0 (the environment).

The loss of residue with time is:

$$\frac{dQ}{dt} = -k_{01}Q \qquad [4.1]$$

which, when integrated, becomes:

$$Q = Q_0 e^{-k_{01}t} \qquad [4.2]$$

where Q is the initial concentration of residue and t is the time elapsed.

From equation 4.2, taking logs:

$$\log Q = \log Q_0 - 0.4343\ k_{01}t \qquad [4.3]$$

If the unicellular organism is suspended in water containing the residue, it will take up residue at R units of mass/unit time. Excretion will still occur and the excretion rate will increase as the internal concentration increases until eventually intake rates will balance loss rates and

$$Qk_{01} = R_{10} \qquad [4.4]$$

The rate of change of residue in the organism:

$$\frac{dQ}{dt} = R_{10} - k_{01}Q \qquad [4.5]$$

Integration gives

$$Q = \frac{R_{10}}{k_{01}}(1 - e^{-k_{01}t}) + Q_0 e^{-k_{01}t} \qquad [4.6]$$

where t is the time interval since exposure began and Q_0 is the amount of residue present when exposure began. If $Q_0 = 0$, then:

$$Q = \frac{R_{10}}{k_{01}}(1 - e^{-k_{01}t}) \qquad [4.7]$$

The exponential decreases as exposure continues and at $t \rightarrow \infty$,

$e^{-k_{01}t} \to 0$, so that

$$Q = \frac{R_{10}}{k_{01}} = Q\infty \qquad [4.8]$$

which is the steady state.

Metabolism of pollutants as well as excretion usually occurs, but this does not alter the example greatly, provided that both the rate of metabolism and the rate of excretion are proportional to the amount of residue.

Normally, studies of pollution involve multicellular animals and specific tissues or organs are considered as compartments (Fig. 4.14). In this case compartment 1 is the blood, which is entered via the food or from the physical environment and residues transfer between other compartments via the blood. Metabolism occurs in compartment 2 (in Moriarty's vertebrate example the liver). Moriarty has drawn several conclusions from compartment models. Firstly, residues do not increase indefinitely with a constant, chronic exposure but a steady state is eventually reached when the amounts of residues remain constant. Secondly, as residues are a function of both uptake and elimination (Fig. 4.13), it cannot be concluded that the levels in a particular organism will depend on its position within the food chain and this may be especially so when direct uptake from the environment is more important than uptake from the food, as is so often the case with aquatic organisms.

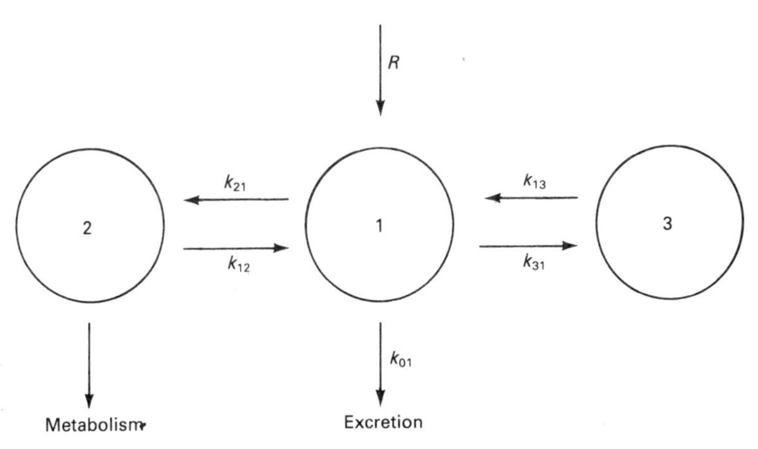

Fig. 4.14. A model for the distribution of DDT within an organism consisting of three compartments. Compartment 1 (blood) absorbs DDT at a steady rate R, while most metabolism occurs in compartment 2 (liver). The rate constants of transfer between compartments is represented by k.

Moriarty (1975a) imagined two species, P_1 and P_2, with individuals having reached steady state concentrations of DDT in their tissues of mean concentration C_1 and C_2 ppm. Individuals of species P_2 (weight W_2) feed only on individuals of P_1, eating x per cent of their body weight per day and eliminating y per cent of the DDT residues per day. To maintain the steady state, the daily uptake of DDT by P_2 must equal the daily loss, so

$$xC_1W_2 = yC_2W_2 \qquad [4.9]$$

or

$$\frac{C_2}{C_1} = \frac{x}{y} \qquad [4.10]$$

Predators will have higher residues than their prey if their food intake, as a percentage of their body weight, is greater than the turnover of their residues, as a percentage of the residues present. If turnover exceeds intake, the residues in the predator will be lower than in the prey.

The loading of an organism with a persistent pollutant will depend on the period of exposure. As predators tend to be larger and longer lived than their prey, they will tend to have heavier loadings. They may also tend to select more heavily contaminated prey, because they will be less fit and easier to catch. Predators may also be less efficient at detoxicating pollutants. These reasons will account for the build up along a food chain. In a Finnish lake, for example, the average DDT content of the plants was 0.5 $\mu g\ kg^{-1}$, of plankton 6.0 $\mu g\ kg^{-1}$, of zoobenthos 14 $\mu g\ kg^{-1}$, of fish 7–42 $\mu g\ kg^{-1}$ and of birds 144–8262 $\mu g\ kg^{-1}$ (Särkkä *et al*, 1978).

Moriarty (1975a) observed that residues of DDT are lost in successively slower phases after exposure, this exponential decrease meaning that small amounts of residue will be exceedingly persistent.

Southworth *et al* (1978) have used a two compartment model (water and *Daphnia*) to examine the bio-accumulation potential of polycyclic aromatic hydrocarbons (PAH), which are released from the combustion of fossil fuels and in some industrial processes and are known to be carcinogenic. They assumed that uptake was a first-order process with respect to PAH concentrations in *Daphnia*. The model was:

$$\frac{dY(t)}{dt} = CZ(t) - kY(t) \qquad [4.11]$$

where $Z(t)$ was the aqueous PAH concentration at time t after the start of the experiment, $Y(t)$ was the *Daphnia* PAH concentration at time t after the start of the experiment, C was the uptake rate constant (h^{-1}) and k was the *Daphnia* elimination rate constant (h^{-1}).

Removal and degradation of PAH due to unknown causes resulted in a decrease in aqueous PAH concentration with time, so that $Z(t)$ was approximated as:

$$Z(t) = \alpha + \beta e^{-\lambda t} \qquad [4.12]$$

If equation 4.12 is substituted into 4.11 and solved:

$$Y(t) = \left[\frac{C\beta}{\lambda - k} - \frac{C\alpha}{k}\right]\exp(-kt) - \frac{C\beta}{\lambda - k}\exp(-\lambda t) + \frac{C\alpha}{k} \qquad [14.13]$$

If the aqueous concentration remains constant, i.e. $Z(t) = \alpha = Z(o)$, $\beta = o$ and $\lambda = \alpha$, then:

$$Y(t) = \frac{CZ(o)}{k}[1 - \exp(-kt)] \qquad [4.14]$$

The parameters C and k were obtained by using an interactive least squares technique. The bioaccumulation curve (concentration factor against time) was obtained from the ratio of equations 4.13 and 4.12 and the results for three representative PAH are given in Fig. 4.15. Both the elimination rate and equilibrium concentration factor were closely related to the chemical structure of the PAH, increasing

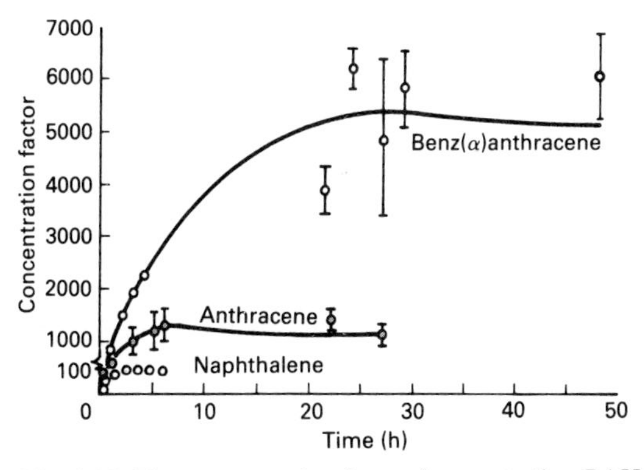

Fig. 4.15. The concentration factor (concentration PAH in *Daphnia*, wet weight/concentration in water) ± 1 S.E. for three PAH at 25 °C (from Southworth *et al*, 1978).

sharply with the increasing molecular weight. PAH were concentrated from 100-fold for naphthalene to 10 000-fold for benz(α)anthracene.

Field studies of toxic pollution

The movement of persistent toxic pollutants through an ecosystem is obviously complex and a study of the whole system is usually impracticable. The prime concern of man is usually the effect of pollutants on himself (Chapter 1) and it is found in practice that only one or two pathways are important, so called *critical pathways*. If pollution is controlled along critical pathways, then adequate control can usually be ensured along other pathways. Critical pathways can of course be developed for target organisms other than man. A generalized critical pathway model is shown in Fig. 4.16. The critical pathways and critical materials (those materials receiving or leading to the greatest degree of exposure) are identified after carrying out a study, usually a desk study, of the various pathways. The limiting environmental capacity is then set, by comparison with primary exposure standards and on the basis of concentration of critical materials per unit rate of introduction. The critical material may be the receiving medium or

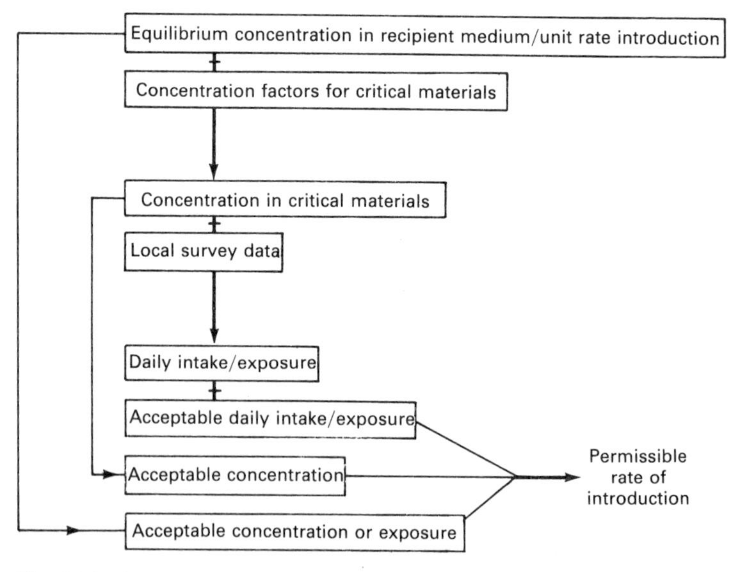

Fig. 4.16. The critical pathway approach for the assessment of toxic effluents (from Preston, 1974).

some living component of the ecosystem which is affected by the pollutant or is important in man's diet.

Toxic pollutants may affect ecosystems by killing out populations of organisms or by reducing their fitness, though sub-lethal effects have received little attention in field situations. The loss of one group of organisms can have serious repercussions on other groups. Toxic pollutants in sewage effluents, for example, may destroy those bacteria responsible for the biodegradation of organic matter, with the result that the oxygen sag curve may extend considerably further downstream than would otherwise be the case.

Solbé (1973, 1977) has made a study of the fish and invertebrates in an English Midland stream, Willow Brook, polluted with zinc from a steel works near its source, as well as with sewage effluent. The maximum observed concentration of zinc was 25 mg Zn l^{-1}, equivalent to more than five times the 48-h LC_{50} of zinc to rainbow trout. The concentration of zinc downstream from the discharge is shown in Fig. 4.17 and the number and biomass of fish species in Figs. 4.18 and 4.19. No fish were found immediately below the discharge and the first species to appear was the stickleback (*Gasterosteus aculeatus*). Both numbers of species and biomass increased gradually downstream. After 1971, the concentration of zinc entering the stream was reduced and the diversity of fish increased in the middle reaches, illustrating the earlier constraining effects of zinc on the fish populations. There were few invertebrates immediately below the discharge of zinc (Fig. 4.19), though tubificid worms and chironomid larvae were abundant. Mayflies, stoneflies and the amphipod *Gammarus pulex* were absent from Willow Brook and zinc

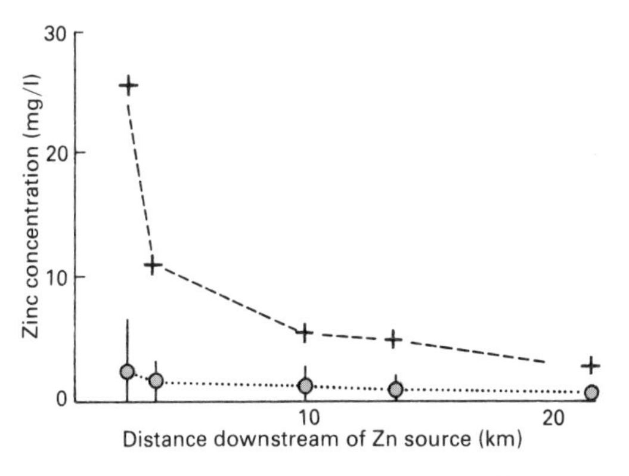

Fig. 4.17. The maximum (+) and average (●) concentrations of zinc (with standard deviations) downstream of a zinc source on Willow Brook (after Solbé, 1977).

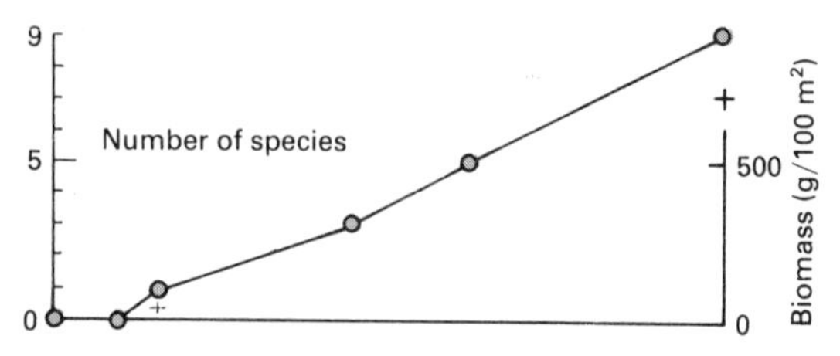

Fig. 4.18. The number of fish species (●) and biomass of fish (+) at stations downstream of a zinc source on Willow Brook (after Solbé, 1977).

concentrations lower than that recorded were found to be toxic to the latter species under laboratory conditions. The diversity of invertebrates increased further downstream.

Some species which are very sensitive to organic pollution, such as stoneflies and caseless caddis, are tolerant of heavy metal pollution. Species diversity in metal polluted waters is reduced, but tolerant species can be very abundant.

Rivers frequently receive complex mixtures of poisons and the five most commonly occurring in English rivers are ammonia, copper, cyanide, phenol and zinc and these may render rivers fishless for many miles. Brown *et al* (1970) have examined several fishless rivers

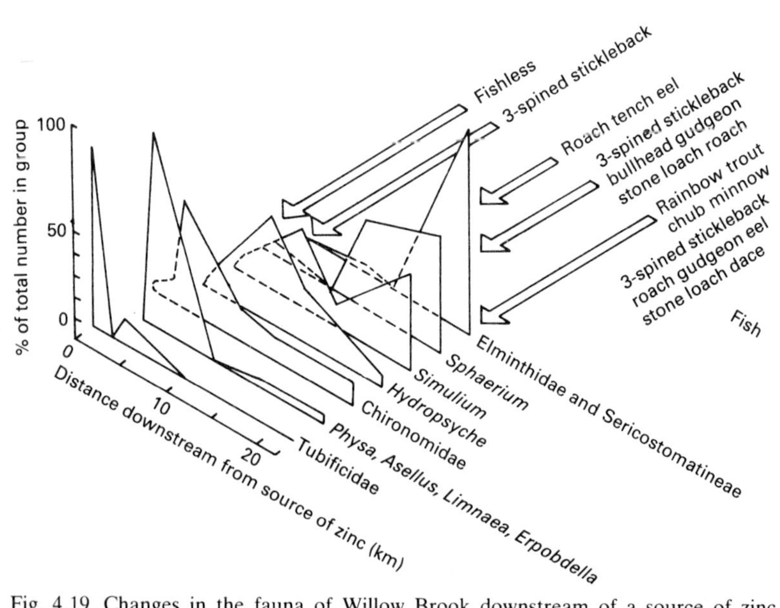

Fig. 4.19. Changes in the fauna of Willow Brook downstream of a source of zinc pollution (after Solbé, 1977).

and determined the acute, lethal toxicity to fish species in the field. Initial observations on the toxicity of river water were made by immersing fish in stainless steel cages in the river. The approximate 48-h LC_{50} of the polluted water was then determined in perspex aquaria on the riverbank by making up a series of dilutions with clean water. The levels of poisons in the rivers were measured and the contribution made by each poison (expressed as a proportion of its 48-h LC_{50}) to the toxicity of the river water was determined from its average concentration in the test solutions up to the time of death of the median fish. These contributions were then summed to describe the overall toxicity of the water. In one river, the Erewash in the English Midlands, ammonia accounted for 40 per cent of the toxicity, though high concentrations of hydrogen cyanide occurred in irregular pulses. In the initial tests with caged fish, all trout died within $2\frac{1}{2}$ h and all roach within 11 h. Toxicity tests were then carried out in 100 per cent, 70 per cent and 50 per cent river water and the occurrence of fish deaths in 70 per cent Erewash water over a 4-day period is shown in Fig. 4.20. There was a high mortality in the tests of both brown and rainbow trout, but little mortality of roach. The increased mortality at 12 noon on 28 October was caused by a pulse of hydrogen cyanide. The concentration of Erewash water killing 50 per cent of the fish in 48 h was 58 per cent for rainbow trout and 85 per cent for the more tolerant roach. It was considered by Brown *et al* that the hydrogen cyanide would have to be eliminated from the river for fish to survive.

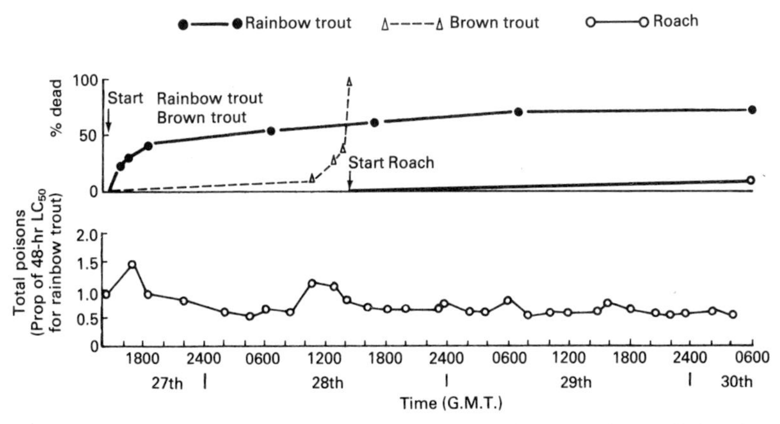

Fig. 4.20. The occurrence of fish deaths in 70 per cent River Erewash water in October 1974, together with the total concentration of poisons, measured as proportion of 48-h LC_{50} for rainbow trout (from Brown *et al*, 1970).

The above technique involved the placing of sensitive organisms into water containing pollutants, whereas an alternative approach would be to examine the perturbation of an ecosystem by adding pollutants. Krebs and Valiella (1978) have investigated aspects of the colonization by fiddler crabs *Uca pugnax* of salt marsh plots fertilized with sewage containing chlorinated hydrocarbon contaminants. Treatment with pesticides reduced the densities of both adult and juvenile crabs and the greatest reduction (86%) occurred at the highest doses. Crabs immigrating into the experimental areas were also killed by the pesticides. As well as causing direct mortality the pesticides considerably slowed down the escape responses of the crabs (Krebs *et al*, 1974), making them more vulnerable to predation (Ward *et al*, 1976) and we can see how this may lead to accumulation along a food chain, as discussed above.

The introduction to this chapter mentioned that biocides are deliberately applied to watercourses to control undesirable animals and plants. Some of these poisons may be very specific, but the majority directly affect many more organisms than the target species, while resultant changes in the structure of food webs have still wider ramifications. Herbicides are particularly widely used for the control of water plants which may impede the flow of water during the summer, when sudden heavy rain can cause flooding. The direct effect of herbicide addition is the loss of macrophytes and non-target organisms such as sensitive invertebrates and fish. Indirect effects include the death and decay of plants, with resultant changes in water chemistry and oxygen levels, the loss of habitat and the loss of food supply (Brooker and Edwards, 1975). Aquatic plants are often replaced by algae. The short term ecological effects are illustrated in Fig. 4.21. Although herbicides may be acutely toxic to fish and invertebrates in laboratory studies, the concentrations normally used in the field are often too low to cause problems. The application of herbicides results in a rapid reduction in photosynthesis, with a consequent reduction in the CO_2 uptake and oxygen output. Brooker and Edwards (1973) reported a decline in pH of the water. The decomposition of large quantities of plant material may cause severe de-oxygenation of the water, resulting in fish-kills, while the release of nutrients during decomposition may stimulate the growth of algae, causing blooms.

Benthic invertebrates may increase in numbers following a herbicide application because of the increased detritus available as food, but those animals closely associated with macrophytes, such as molluscs, caddis and some chironomids, suffer a severe reduction in

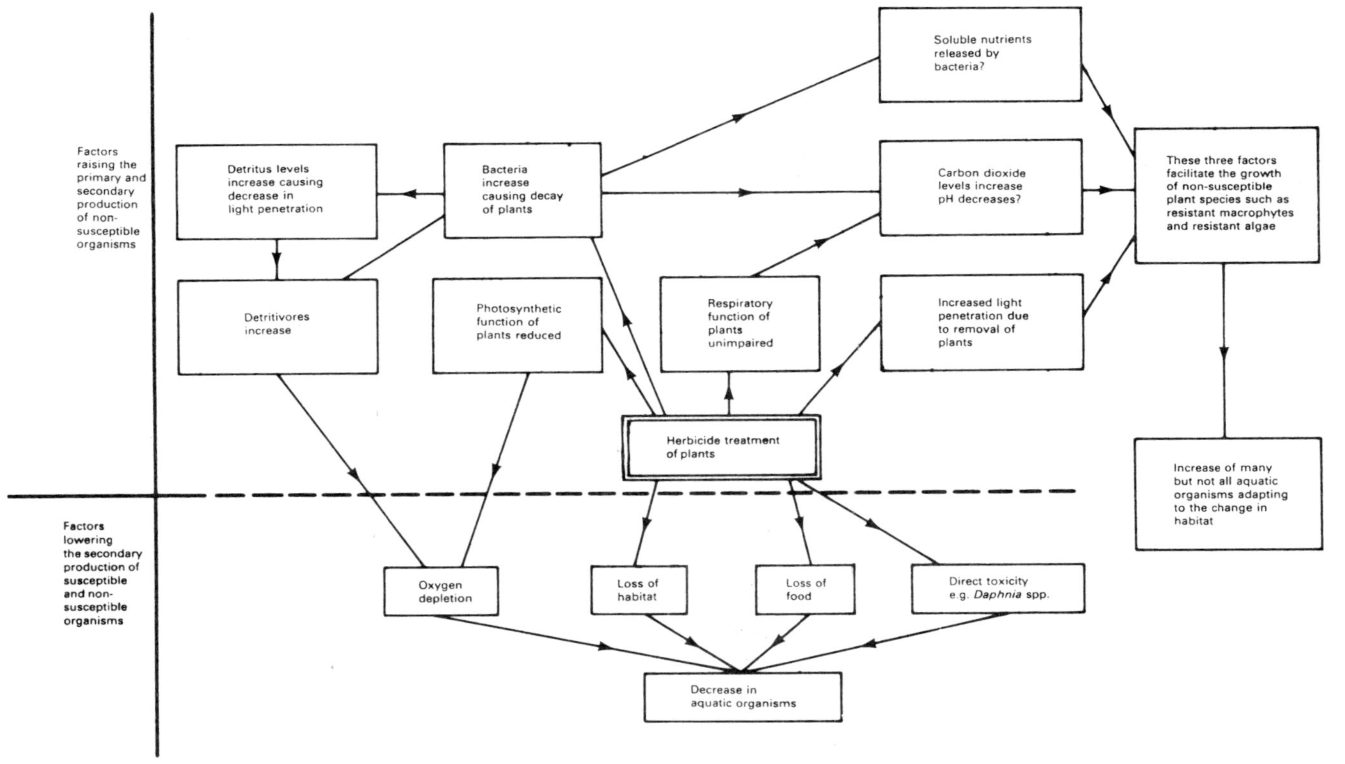

Fig. 4.21. The possible short-term ecological effects of herbicides in aquatic ecosystems (from Newbold, 1975).

numbers, which may still be apparent in the following year after the plants have recovered (Brooker and Edwards, 1974). The change in invertebrates may affect the diet of predators and could eliminate specialists. Brooker and Edwards (1974) found that the diet of eels (*Anguilla anguilla*) changed, after herbicide application to a reservoir, with benthic chironomids being of considerably increased importance.

Chapter 5

THERMAL AND MINE POLLUTION

Introduction

The last three chapters have dealt with the major types of pollution affecting freshwaters. Other pollutants may cause particular problems or may be particularly insidious. Radionuclides, for example, emanating chiefly from nuclear power stations and fall-out from weapons testing behave similarly to toxic pollutants but their potentially powerful carcinogenic and genetic effects and great persistence in ecosystems has been widely recognized. Their release into the environment is strictly monitored and controlled (e.g. Nelson *et al*, 1972; Preston, 1972), though far reaching pollution disasters may result from accidents at power stations or from nuclear wars. Oil is causing increasing problems in freshwaters, though again the effects are largely toxic. Suspended solids released into water courses may be either inorganic (e.g. from mineral workings, logging operations) or organic but the effects of both on the substratum and on the organisms of rivers are similar and are discussed in Chapter 2.

Thermal pollution and pollution arising from mine drainage is very widespread and deserves special consideration.

Thermal pollution

Electricity generating stations require large amounts of cooling water and this is abstracted and returned to rivers, estuaries or the sea. The water is returned at a higher temperature than it is abstracted and this waste heat can potentially cause serious ecological problems in the receiving water. Power stations have to dispose of some 50–65 per cent of the energy they generate as heat to the environment via the cooling water (Langford, 1972). In 1975, 59.4×10^6 m^3 d^{-1} of cooling water was discharged into rivers in England and Wales, 93 per cent from electricity generating stations (Department of the Environment, 1978). Langford (1972) estimated that 40 per cent of the mean annual run-off of water in Britain is circulated in power stations, while the figure for the USA is 10 per cent. Most of this water can be re-used and most of it is free of contamination other than heat, though there may be reduced levels of ammonia and organic nitrogen and increased levels of nitrate, chlorine and suspended solids compared with water in the power station intake.

Langford (1970) has studied in detail the effects of a power station at Ironbridge, in west midland England, on the temperature of the River Severn. The natural annual range of temperature was increased by up to 6 °C (e.g. the maximum temperature in 1970 400 m above the power station intake was 22 °C, while 2000 m below the outfall it was 28 °C). The daily increments downstream ranged from 0.5 °C during spates to 7.2 °C in low flow conditions and the diurnal variation in summer was increased by more than 100 per cent. In spring the rising mean temperature was advanced by 3–4 weeks, while the fall in temperature in autumn was delayed by 1–3 weeks.

There are a number of possible effects of heated effluents on the biology of receiving waters. Those species intolerant of warm conditions may disappear, while other species not normally found in unheated water may thrive, e.g. the tubificid worm *Branchiura sowerbyi*, so that the structure of the community may change. Respiration and growth rates may be changed and these may alter the feeding rates of organisms. The reproductive period may be brought forward and development may be speeded up. Parasites and diseases may also be influenced. Indirect effects of thermal pollution may include a reduction in the oxygen concentration of the water, particularly when organic pollution is also present, resulting in the loss of sensitive species. Increased temperatures may also render organisms more vulnerable to the effects of toxic pollutants present in the water.

Many species are able to acclimate to the normal range of temperatures occurring below thermal discharges and large scale mortalities occur only when populations are trapped in an effluent channel or when sudden discharges of hot water occur. Some fish populations are unaffected by rapid temperature fluctuations of 12–15 °C (Langford and Aston, 1972), but these temperature changes may be too great for cold water fish. McFarlane *et al* (1976) have shown that three cyprinid minnow species which inhabit the tributary streams of the Savannah River in USA., but do not occur in adjacent thermally stressed streams, survived well in temperatures up to 32 °C, but died rapidly at 32–34 °C. Figure 5.1 illustrates the rates of survival of the

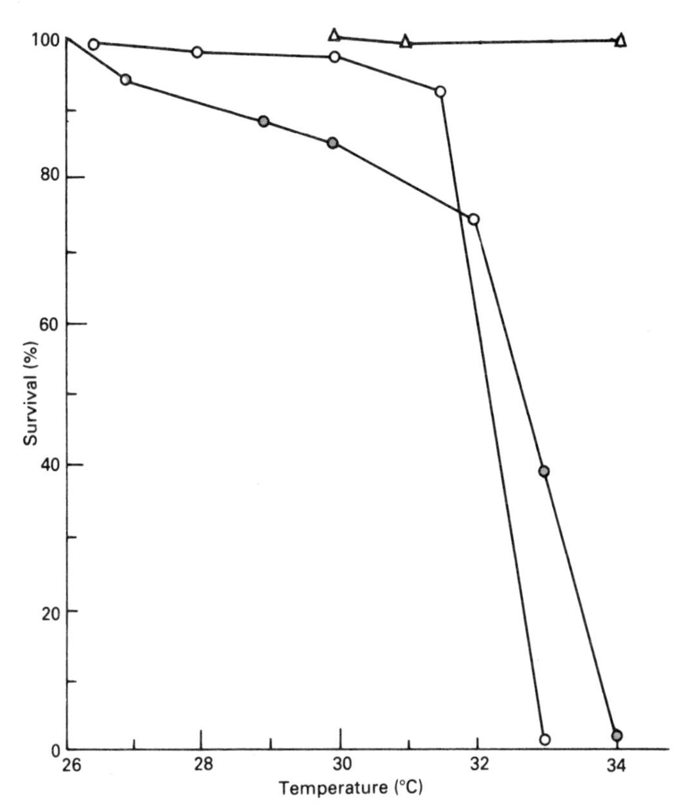

Fig. 5.1. The survival of groups of the yellowfin shiner *Notropis lutipennis* at three rates of temperature increase. ●–● 4 °C h^{-1}, ○–○ 1 °C h^{-1}, △–△ 0.125 °C h^{-1} (after McFarlane *et al*, 1976).

yellowfin shiner *Notropis lutipennis* at different rates of temperature increase. A heating rate of 4 °C h^{-1} gave a 75 per cent survival at 32 °C but a 100 per cent mortality in less than one hour at 34 °C. With an increase in rate of 1 °C h^{-1}, 92 per cent survived at 32 °C but the fish died very quickly at 33 °C, whilst at a heating rate of 0.125 °C h^{-1} survival was excellent to 34 °C, but at this temperature all fish died overnight. Alabaster (1962) has shown that, with temperatures up to 26 °C below thermal discharges fish tend to be attracted, whereas at temperatures above 30 °C they tend to move away. Anglers certainly make use of the attractive effect of heated discharges to fishes.

The effects of a thermal effluent on the growth of an organism will depend on the relative apportionment of energy into respiration and growth. If the increase in temperature results in increased metabolism, with no increase in feeding rate, then less energy may be diverted to growth. However, the increased temperature may extend the period over which growth can occur, bringing forward the onset of growth in spring and extending the growing period into the autumn. Mobile species may avoid periods when discharges are resulting in higher temperatures than the species' optimum so that fishes living in thermally polluted areas will not show significantly different rates of growth to fishes in neighbouring unpolluted waters (Spigarelli and Smith, 1976). Other studies have shown an increase both in growth rate of fish and in the maximum size attained (e.g. White *et al*, 1976).

Temperature changes may increase the vulnerability of a species to predation and parasitism. Coutant *et al* (1976) observed that rapid temperature decreases of about 6 °C, which may occur when the quantity of thermal effluent is reduced, increase the susceptibility of juvenile channel catfish (*Ictalurus punctatus*) to predation by large-mouth bass (*Micropterus salmoides*). Young bluegills (*Lepomis macrochirus*) also became significantly more vulnerable to predation by large-mouth bass when exposed to a 9 °C cold-shock (Wolters and Coutant, 1976). With reference to parasites Aho *et al* (1976) found that, with the mosquito fish *Gambusia affinis*, the metacercariae of the brain parasite *Ornithodiplostomum ptychocheilus* occurred at higher densities in fish from thermally polluted waters, whereas the metacercariae of the body-cavity trematode *Diplostomum scheuringi* were most numerous in fish from waters not receiving thermal pollution. These differences may have been due to the effects of temperature on the life cycles of the parasites or to effects on either the intermediate or definitivc hosts.

The life-histories of organisms may be altered by thermal pollution, for instance by accelerating development time, and some fish spawn earlier in power station effluents than they do in neighbouring unheated waters (Langford, 1972). However, the emergence of insects from the River Severn, discussed above, was not noticeably altered by heated water and natural variability due to other environmental factors masked any effect due to temperature (Langford, 1975).

Cherry *et al* (1974) examined the temperature influences on bacterial populations and found the highest diversity in bacterial types to be present in the temperature range 16–19 °C, temperatures above this decreasing diversity but increasing the total population. At lower temperatures both diversity and populations were reduced. Oden (1979) observed that the population density of the littoral meiofauna in a reservoir was reduced at sites receiving thermal effluents and a detailed analysis of the rotifer community showed a decreased species richness compared with a site not receiving effluents. Hillbricht-Ilkowska and Zdanowski (1978) found a decrease in the planktonic rotifer populations in thermally polluted lakes. The copepods also decreased, but at the same time cladoceran crustaceans increased and these increased the circulation of materials within the lakes.

Except in unusually severe thermally polluted conditions (e.g. Coutant, 1962), it appears that macro-invertebrate communities of rivers are relatively little affected by thermal discharges (Langford and Howells, 1977) and generally the same is true of fish communities (e.g. Teppen and Gammon, 1976), fish having the ability to vacate water which is temporarily inimicable to them (McFarlane, 1976). Overall it can be concluded that thermal pollution has not been as damaging to aquatic ecosystems as was originally feared and, within the usual range of temperature increases caused by power stations, the homeostatic mechanisms at work within the community minimize the damage.

Recent efforts have been made to put the waste heat in cooling water to biological use. This has included growing both marine fish (e.g. soles *Solea solea* and freshwater fish (e.g. eels *Anguilla anguilla* and carp *Cyprinus carpio*) in cooling water and greatly increased growth rates have been achieved. Greenhouses have also been heated with cooling water at substantially less cost than other heating systems. Attempts are also being made to link power station operations with sewage treatment works. Mixing of warm power station effluent with raw sewage accelerates the treatment process, whilst the circulation of partially treated sewage effluent around the cooling tower accelerates the conversion of organic nitrogenous compounds and ammonia to nitrites and nitrates.

Acid mine drainage

Drainage water from mines causes very considerable pollution problems over wide areas. Lundgren *et al* (1972) report that about 16 000 km of streams and 11 700 ha of impoundments and reservoirs are seriously affected by pollution from surface mining (strip or open cast) operations in USA, while deep mining adds considerably to these figures. In England and Wales in 1975, 366 discharges released 881 990 m^3 d^{-1} of minewater into rivers and canals (Department of the Environment, 1978). Coal mines are the single most important cause of this pollution, though the mining of other ores causes local water quality problems. Abandoned mines are more of a problem than working mines. When coal is mined, substantial quantities of the mineral pyrite, a crystal composed of reduced iron and sulphur (FeS_2), are often exposed to the oxidizing action of air, water and chemosynthetic bacteria and these utilize the energy obtained from the conversion of the inorganic sulphur to sulphate and sulphuric acid. The reactions are as follows:

$$2FeS_2 + 2H_2O + 7O_2 \longrightarrow 2FeSO_4 + 2H_2SO_4$$
$$2FeSO_4 + O_2 + 2H_2SO_4 \longrightarrow 2Fe(SO_4)_3 + 2H_2O$$
$$Fe(SO_4)_3 + 6H_2O \longrightarrow 2Fe(OH)_3 + 3H_2SO_4$$

At higher pH pyrite can be oxidized in the presence of ferric iron:

$$FeS_2 + 14Fe^{3+} + 8H_2O \longrightarrow 15Fe^2 + 2SO_4{}^{2-} + 16H^+$$

The conversion of ferrous sulphate to ferric sulphate occurs very slowly below pH4, but it is rapid in the presence of microbial catalysts such as iron-oxidizing bacteria (Lundgren *et al*, 1972). The mechanisms of pyrite oxidation are discussed in more detail in Dugan (1972).

The most numerous bacteria in acid mine waters are those which oxidize sulphur (*Thiobacillus thiooxidans*) and iron (*Thiobacillus ferrooxidans*) and their densities may be as high as 10^9 ml^{-1}. Walsh (1978) reported that iron-oxidizing *Metallogenium* are active in mine waters of between pH 3.5 and 4.5 and create the lower acidic conditions (below pH 3.5) suitable for the functioning of *Thiobacillus ferrooxidans*. The biological catalysis of the reaction is unnecessary at pH greater than 4.5.

The types of water issuing from mines can be divided into four classes (Table 5.1), depending on the kind of mine and the surrounding geology. The drainage of coal mines is dependent on the quantities of sulphides and their spatial distribution, crystallinity and size as well as the presence of bacteria, the water level and the presence of calcium in the sulphide aggregates (Lundgren *et al*, 1972).

The high acidity of mine water also brings heavy metals into solution, adding to the pollution problems. Those organisms in the receiving stream responsible for breaking down organic matter may be destroyed by the high acidity so that the self-purification process is inhibited and oxygen depleted water, often with suspended solids, extends much further downstream than would otherwise be the case. Washings from active mines also result in a heavy load of suspended matter. As the pH increases downstream, due to the dilution of mine water with stream water and run-off, ferric hydroxide precipitates out

Table 5.1 Classes of mine drainage (from Lundgren *et al* 1972).

	Class I, acid discharges	*Class II, partially oxidized and/or neutralized*	*Class III, oxidized and neutralized and/or alkaline*	*Class IV, neutralized and not oxidized*
pH	*2–4.5*	*3.5–6.6*	*6.5–8.5*	*6.5–8.5*
Acidity, mg/l ($CaCO_3$)	1 000–15 000	0–1 000	0	0
Ferrous iron, mg/l	500–10 000	0–500	0	50–1 000
Ferric iron, mg/l	0	0–1 000	0	0
Aluminium, mg/l	0–2 000	0–20	0	0
Sulphate, mg/l	1 000–20 000	500–10 000	500–10 000	500–10 000

on to the stream bottom, covering the substratum with a brown slime and smothering benthic algae and macrophytes.

Hargreaves *et al* (1975) have described the photosynthetic flora from fifteen sites, thirteen associated with coal mines, having a pH of less than 3.0 in England. All of the sites also had high levels of one or more heavy metals and of silica, while most had high levels of phosphate, ammonia and nitrate. Twenty-four species of photosynthetic plants were found in flowing waters, with an additional four species restricted to pool sites. *Euglena mutabilis* proved to be the most widespread species and was also the most abundant, sometimes forming 80 per cent cover. Some species, although widespread, could not survive the lowest pH and species richness was greatest in the range pH 2.5–3.0. Four species were found to have been recorded both in the English survey and in studies of acid streams in the USA, namely *Euglena mutabilis*, *Lepocinclis ovum*, *Eunotia exigua* and *Ulothrix zonata*.

A number of studies have been made of the effects of acid mine drainage on the communities of invertebrates in streams. Letterman and Mitsch (1978) examined Ben's Creek, in Pennsylvania, USA, where the inputs of mine drainage entering over the past 50 years could be considered as a point source. Five of the sampling stations were above the discharge, four were in the area where the discharge entered and a further six were below the discharge. The entry of alkaline water ensured that pH was not affected by the drainage, but iron, conductivity, hardness and sulphate all increased significantly, with sulphate increasing from 5–10 mg l^{-1} above the discharge to 100–300 mg l^{-1} below. The abundance of the major groups of

Table 5.2. The benthic invertebrate community found upstream and downstream of mine drainage in Ben's Creek, Pennsylvania (after Letterman and Mitsch, 1978).

Order	*Upstream stations*			*Downstream stations*			
	1	3	6	9	11	13	15
Trichoptera	63	71	58	2	7	11	58
Ephemeroptera	53	102	45	0	4	21	18
Plecoptera	43	45	8	3	27	3	4
Odonata	19	7	1	0	0	1	1
Coleoptera	11	6	17	1	4	3	2
Diptera	246	74	49	2	21	37	33
Hydracarina	3	3	16	3	0	7	10
Megaloptera	0	0	0	1	1	2	1
Oligochaeta	2	0	21	0	1	0	0
Decapoda	2	4	1	0	2	0	0
Total	442	312	216	12	67	85	127
No. m^2	952	672	465	26	144	183	273

invertebrates above and below the discharge is shown in Table 5.2. The mine drainage eliminated four groups immediately downstream, while the total numbers of invertebrates decreased some twenty-fold. The numbers had not recovered at the lowest station some 9 km downstream. Chironomids were the most tolerant group and the caddis *Hydropsyche* was also considered tolerant, possibly receiving protection from the ferric hydroxide by its silken net used for catching food. The stonefly *Isogenus* was considered semi-tolerant as it reappeared 1 km downstream of the discharge. Dragonfly larvae and the caddis *Rhyacophila* were intolerant. In a stream, receiving mine drainage, with a very low pH 2.6–3.0, Koryak *et al* (1972) found a depauperate community dominated by *Chironomus riparius*, together with a few cranefly larvae (Tipulidae) and, as conditions improved slightly, small numbers of Neuroptera and Coleoptera appeared. The acid water below the discharge resulted in a low diversity of species, but with large populations, released from competition and predation by fish. Further downstream, in the zone of neutralization, the precipitation of iron, which alters the characteristics of the substratum and clogs the gills of invertebrates, resulted in both low diversity and low population density.

Fish are very severely affected by acid mine drainage and low pH. Letterman and Mitsch (1978) recorded ten species of fish, with a biomass of 22 812 kg ha^{-1} upstream of the mine discharge in Ben's Creek. Downstream of the discharge only six species occurred, with a biomass of 11 kg ha^{-1}. The overall populations above and below the discharge were 31 056 ha^{-1} and 963 ha^{-1} respectively.

If the source of pollution is water percolating through mine waste heaps, it can often be treated successfully by covering the tip with clay and top soil. Water coming from springs or underground water sources tends to be more acidic, with higher loads of metals, and is much more difficult to control. It can be neutralized with lime (Walsh, 1978) but this is very expensive and frequently ineffective. The succession of *Metallogenium* to *Thiobacillus ferrooxidans* which occurs as the pH drops may provide a new, biological approach to preventing pollution. If the onset of *T. ferrooxidans* activity, which results in an exponential increase over a limited time period of the production of ferric iron, could be delayed by a small percentage increase of the mine water residence time, the majority of the highly polluting later cycles of the pyrite degradation cycle could be avoided. Walsh and Mitchell (1975) have demonstrated in the laboratory that acidity and iron release can be controlled by perfusing mine water with artificial mine water enriched with ferrous iron, which inhibits *Metallogenium* growth. The inhibition of *Metallogenium* causes a reduction in the activity of later members in the succession, such as *T. ferrooxidans*.

In conclusion, brief mention needs to be made of the phenomenon of acid rain, its effect on the pH of freshwaters being similar to that of mine drainage. Aerial pollutants, particularly sulphur dioxide, are deposited in rainfall, often far from their origins and this acid rain can sharply lower the pH of the receiving stream or lakes, being particularly detrimental to fish communities. The problem has received especial attention in Scandinavia, which receives airborne pollutants from industrial nations further south.

Chapter 6

THE BIOLOGICAL ASSESSMENT OF WATER QUALITY IN THE FIELD

Introduction

The biological assessment of water quality in the field may involve a number of levels of effort; survey; surveillance; monitoring or research (p. 9) and the objectives of a particular study must be clearly defined before the work programme is begun. A research worker will usually study a particular problem in depth, but the biologist working within the water industry will mostly be involved in large scale survey, surveillance or monitoring studies, necessitating visits to many, often widely-scattered, localities during the course of a year, frequently for different purposes. Severe constraints will be placed on the amount of sampling that can be undertaken at any particular site and a compromise will have to be reached between the level of sampling effort and the amount of data required to produce meaningful results. It will not be feasible to sample the entire biota so that a decision must be made as to which group of organisms will provide the most information for solving a particular problem.

Ideally a sample should be compared with detailed information from past samples for that site, but frequently the biologist is involved only after a pollution incident has occurred. It may be possible to find an acceptable control site, for example upstream of the incident, or in a local tributary, but the assumption then has to be made that the biological community was identical in both the unpolluted control and the polluted site before the incident took place. Mason and Bryant (1975) surveyed twenty-eight shallow lakes in the Broadland area of eastern England and many of them were found to have a depauperate flora and fauna and were suffering severe eutrophication (p. 77). Despite the fact that the area had been renowned for decades for its biological diversity and importance for conservation, no previous survey had been carried out so that the extent of the ecological degradation had to be pieced together from published natural history notes, unpublished reports and the personal recollections of local naturalists. In no cases were the survey methods of previous observations recorded. Of these twenty-eight sites, *three* were still biologically rich and could be considered as possible controls, but it was known that in *two* of these lakes marked changes had occurred (severe pH fluctuations in one and changes in the dominance of macrophytes in the other). However, it was not known whether these changes were due to natural factors or anthropogenic influences. The biologist investigating pollution is often faced with these dilemmas.

The ecology and sensitivity to pollution of many of the organisms used in water quality assessments is still very inadequately known, even in those countries with a long tradition of research in freshwater biology. The key organisms in the economy of one particular watershed may be insignificant in a neighbouring watershed of differing geology or flow regime. Pollutants affect organisms differentially at different life stages or different times of year, while natural changes in population size may underlie any of the effects of pollution, so that the disentanglement of these from natural events is often very difficult.

The research biologist communicates his findings chiefly to other scientists and, providing his study has been carefully planned and executed, this is relatively easy. A biologist in the water industry has the task of explaining his findings to managers, who are usually not biologists and often not scientists, or to the general public. The biologist therefore frequently resorts to deriving indices of water quality, single numbers which may help in communication but which result in a considerable loss of biological information. The danger is that the biologist himself may come to rely too heavily on interpreting change in terms of indices, such that more subtle analyses of the data, giving increased biological understanding, will not be attempted. Indeed a biologist in the water industry may not be allowed time by his managers to undertake more penetrating approaches to the understanding of pollution problems.

Despite these many difficulties, biological assessments of water quality have proved very successful and even mild, intermittent pollution, frequently missed by routine chemical sampling, has been detected. This chapter will deal chiefly with survey and monitoring techniques, rather than with the undertaking of specific research projects. The monitoring of intrusive organisms, such as faecal bacteria, and bioassay methods will be reserved for discussion in Chapter 7.

Sampling strategy

The definition of objectives is the essential first stage in the design of a sampling programme. Hellawell (1977) has suggested that there are three prime objectives in surveillance and monitoring studies for water quality. Firstly, there is environmental surveillance, where the objective is to detect and measure adverse environmental changes, such as the effects of unknown or intermittent pollutants, or to follow the improvement of conditions once a pollutant has been removed. The second objective is to establish water quality criteria in which causal relationships between ecological changes and physico-chemical parameters are determined. The third main objective is the appraisal of resources. This may involve a large scale survey to assess general water quality. Alternatively, particular water resource problems, involving nuisance species, or the impact of new developments, may be investigated. These will also include the management of fisheries or the conservation of threatened species.

Sampling design is dealt with by Elliott (1977), Hellawell (1978), Downing (1979), Green (1979) and Jeffers (1979). Green (1979) lists ten principles to be taken into account in the design of a sampling programme. These stress the need for randomly allocated replicates and controls, matters also discussed by Jeffers (1978). Green points out that a preliminary sampling should be undertaken to evaluate the sampling design and examine the options for subsequent statistical analysis and this often saves considerable time later in the programme, by emphasizing design faults at the beginning of the study. The efficiency of the sampling device should be determined at the beginning, as should the size and number of samples in relation to the size, density and spatial distribution of the organisms being sampled.

Surveys may be either extensive or intensive. Extensive surveys aim to discover what species are present in an area, usually with a measure of relative abundance, and are especially used where the water quality over many sites is being monitored or compared. Such surveys have been criticized, or even considered valueless (e.g. Gray, 1976), because they are too superficial to detect or interpret subtle environmental changes, such as alterations in species dominance due to biological interactions, so that it is impossible to disentangle natural changes from those caused by pollution. Such a view is undoubtedly too pessimistic and examples will be given later of the ability of faunal surveys to detect pollution without a detailed foreknowledge of the ecology of a site. It is probably true, however, that sampling design is frequently given inadequate thought in extensive surveys.

Intensive surveys usually aim to determine population densities. Elliott (1977) listed the main considerations in designing a quantitative survey as the dimensions of the sampling unit, the number of sampling units in each sample and the location of sampling units within the sampling area. Populations of organisms are usually highly aggregated so that a large number of samples are frequently required to obtain a population estimate that is statistically meaningful. Chutter and Noble (1966) and Chutter (1972), using benthic macroinvertebrate data from streams in South Africa and California, considered that over fifty replicates are required to attain an estimate of population size to within 20 per cent. For some species, many more samples would be needed. Ninety-eight samples were required to obtain an estimate of mean density (±40 per cent), with 95 per cent confidence limits for the limpet *Ancylus fluviatilis*, for example (Edwards *et al*, 1975). Clearly such sampling intensity would be impossible in an extensive survey and even in an intensive survey the rarer species will be inadequately sampled. Edwards *et al* (1975) have shown that, sampling a riffle, only 44 per cent of the species taken would be common to both of any two random samples.

The number of samples required for a specified degree of precision can be readily calculated if an estimate of the mean abundance is made from a pilot survey:

$$D = \frac{1}{x}\sqrt{\frac{s^2}{n}} \qquad [6.1]$$

where D is the index of precision, $\bar{x}$ is the mean, s^2 the variance and n the number of samples. A standard error of 20 per cent of the mean is usually acceptable for ecological studies so that the number of samples required to obtain this level of precision ($D = 0.2$) is:

$$n = \frac{s^2}{0.2^2 x^2} \qquad [6.2]$$

If a series of samples are taken over time it must be remembered that the number of samples required to maintain this level of precision will change as the population size and degree of aggregation changes. For example, on collecting 30 random benthic samples of the tubificid worm *Potamothrix hammoniensis* in a shallow lake, the standard error with a population density in July of 10 129 m^{-2} was 16 per cent but after the death of adult worms during the late summer the standard error in October increased to 45 per cent of the population mean of 660 m^{-2} (Mason, 1977a).

A small sampling unit is generally preferable to a large unit because more samples can be handled for the same amount of effort and the statistical error in estimating the mean is reduced. Many small units cover a greater range of habitats than a few large units so that the population estimate will be more representative of the sample area.

Sampling units must be selected at random from within the sampling area for the sample to be representative of the whole population. To prevent all of the sampling units falling randomly in one part of the sampling area, or if the location to be sampled shows environmental patten, stratified random sampling is often used. The sampling area is divided into smaller areas (strata), usually of equal size, and sampling units, divided equally among the strata, are selected at random from within each stratum on each sampling occasion.

The choice of organisms for surveillance

It has already been stated that it is usually impossible to study the entire biota present in a sampling area because of the constraints of time and of the wide variety of sampling methods required for different groups of organisms. A survey or monitoring programme must therefore be based on those organisms which are most likely to provide the right information to answer the questions being posed.

The use of a single species as a water quality indicator is generally avoided because individual species show a high degree of temporal and spatial variation due to habitat and biotic factors and these confuse any attempt to relate presence, absence or population level with water quality. Furthermore, considerable care is needed in identification as similar species may show very different reactions to pollution. To take an estuarine example from the polychaete worm family Capitellidae, *Capitella capitata* indicates polluted environments, whereas the closely similar *Capitella ambiesta* occurs in unpolluted conditions (Reish, 1973). For indicator species to be worthwhile, they must be able to be used to detect subtle, rather than gross and obvious, effects of pollution.

The use of communities of organisms allows this more subtle approach. To be suitable for a broad survey or monitoring programme a biological system requires the following features (Price, 1978):

1. the presence or absence of an organism must be a function of water quality rather than of other ecological factors;

2. the system must reliably assess water quality, be reliably expressible in a simplified form, yet be sufficiently quantifiable to allow for comparisons;
3. the assessment should relate to water quality conditions over an extended period, rather than just at the time of sampling;
4. it is frequently important that the assessment should relate to the point of sampling rather than to the watercourse as a whole;
5. sampling, sorting, identification and data processing should be as simple as possible, involving the minimum of time and manpower.

Numerical abundance at some sites, widespread distribution and a well documented ecology are also important factors to take into account in selecting a group of organisms for water quality assessment. Hellawell (1978) lists the advantages and disadvantages of different taxa as indicators of pollution. Fish tend to be too mobile and hence can avoid intermittent pollution incidents, while their capture requires considerable manpower. They are, however, easy to identify, their ecology and physiology is relatively well known and, as they are at the top of the food chain, they may reflect changes in the community as a whole. Macrophytes are seasonal, are tolerant of intermittent pollution and are strongly influenced by geology, soil type and management practices (e.g. for flood prevention), though Haslam (1978) describes in some detail the relationship between water quality and plant distribution. Much information on the tolerance of pollution by algae (e.g. Patrick, 1954; Fjerdingstad, 1965) and protozoa (e.g. Bick, 1968) is available, but considerable time and expertise are required for their enumeration and identification. Algae are popular organisms for the assessment of water quality and are especially valuable in examining the eutrophication of lakes, but Hellawell (1977) found that most workers recommended macro-invertebrates for water quality assessments. The remainder of this chapter will concentrate on benthic macro-invertebrates; methods for other groups are fully reviewed in Hellawell (1978).

Macro-invertebrates and water quality assessment.

Hellawell (1977) lists a number of reasons for preferring benthic macro-invertebrates. The sampling procedures are relatively well developed and can be operated by someone working alone, while there are identification keys for most groups. Macro-invertebrates are reasonably sedentary, with comparatively long lives, so that they can be used to assess water quality at a single site over a long period of time. The group is heterogeneous in that a single sampling technique may catch a considerable number of species from a range of phyla, therefore it is likely that some species or groups will respond to a particular environmental change. Macro-invertebrates are generally abundant.

There are some disadvantages to choosing macro-invertebrates. Their aggregated distribution means that, to obtain a representative sample of a site, many samples need to be taken. The muddy, depositing substrata of the lowland reaches of rivers, or of lakes, are often dominated by chironomids and tubificid worms, which are difficult to identify, while the water in these situations is frequently deep, making sampling difficult. The insect members of the community may be absent for part of the year, so that care needs to be taken in the interpretation of monitoring results.

Sampling for macro-invertebrates

The simplest method of sampling for macro-invertebrates, suitable for shallow waters over eroding substrata, is the kick sample. The operator faces downstream and holds a standard pond net vertically in front of him, with the bottom against the substratum. The substratum upstream of the net is then vigorously disturbed with the feet and the dislodged invertebrates flow into the net. In shallow waters stones can also be turned over by hand in front of the net. By attempting to disturb a known area, or by kicking for a fixed period of time, this method can be made semi-quantitative for relative abundance estimates. The technique is rapid and inexpensive and is particularly suited to faunal surveys or extensive surveillance programmes.

A variety of samplers have been designed for the quantitative collection of invertebrates and these have been reviewed by Hellawell (1978). Two widely used samplers are those of Surber (1937), combining a quadrat with a net (Fig. 6.1a), and cylinder samplers (Fig. 6.1b). The Surber sampler consists of a net with a hinged frame, attached to its lower margin, which can be pushed onto the substratum and locked into place. The frame quadrat encloses 0.09 m^2 (1 sq ft) and stones and gravel within this area are lifted and stirred so that invertebrates are dislodged into the net. Side-wings on the net help to reduce the loss of animals round the net.

In shallow waters, open-ended cylinders can be pushed into the substratum to enclose a known area, a typical example being the Neil sampler (Macan, 1958), which has two openings near the base. The

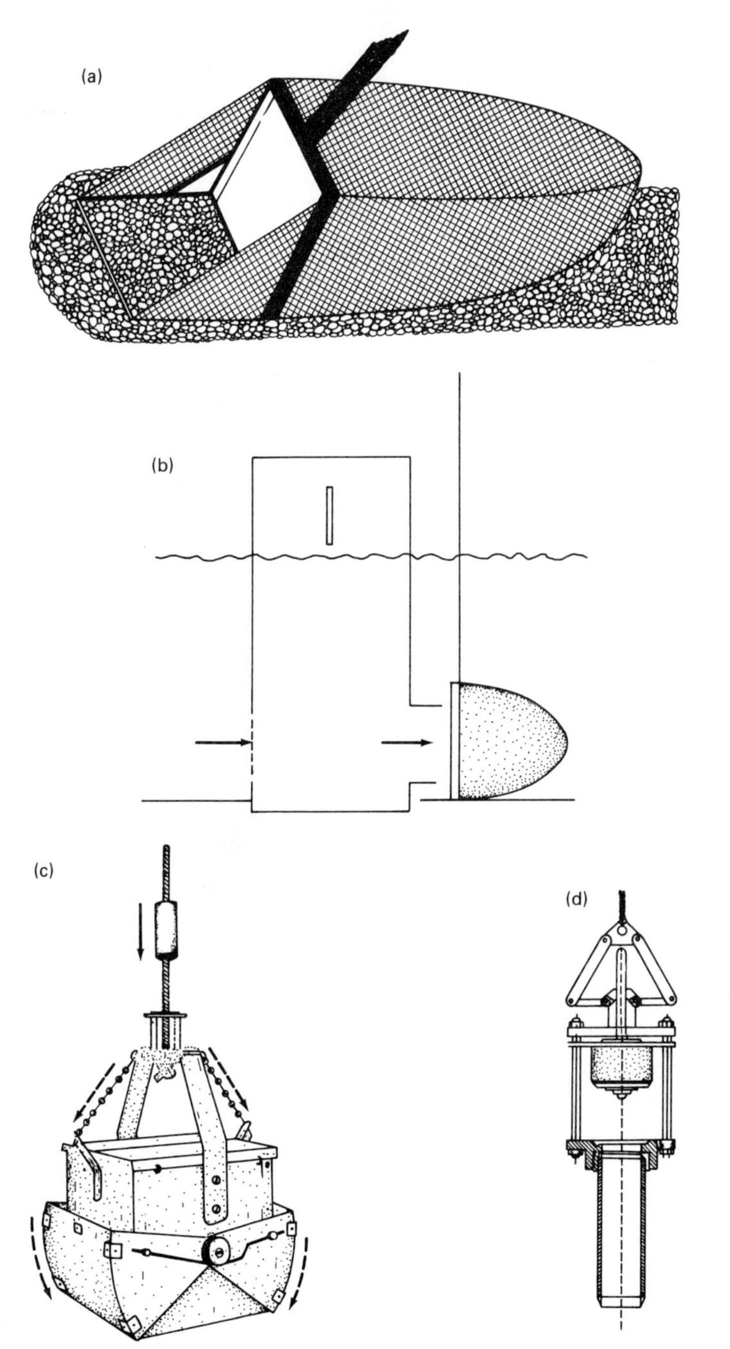

Fig. 6.1. Types of samplers for benthic invertebrates. (a) Surber sampler (b) cylinder sampler (c) Ekman grab (d) corer (a, c, d after Hellawell 1978).

upstream opening is covered with a coarse metal mesh which allows a through-flow of water but excludes drifting invertebrates. The downstream opening has an attached net for collecting animals. The openings of the Neil sampler have sliding doors which are opened after vigorously stirring the enclosed substratum and the sampler is traditionally made of stainless steel or aluminium. A considerably cheaper version can be made by using a 50 cm length of 25 cm diameter plastic sewer piping. The moveable doors can be dispensed with and a standard pond-net can be held over the downstream aperture as the substratum is stirred (Fig. 6.1b)

For shallow, still waters, such as ponds or the muddy edges of reservoirs, a cylinder sampler without inflow and outflow apertures can be used. The sediment may be removed with a plastic beaker or a small, fine-mesh hand net and the author has found this method highly satisfactory for sampling the benthos of shallow, coastal lagoons. In water less than 2 m deep sediment can be collected from a boat using a pond-net and this can be calibrated against a standard sampler (Mason, 1977a), this method resulting in a considerable saving of time. For the sampling of deeper waters of lakes and rivers a variety of grab and core samplers have been devised and these are reviewed by Hellawell (1978). The jaws of grabs (Fig. 6.1c) close beneath a known area of bottom, the mechanism being operated using a messenger released from the surface. For very fine sediments, corers (Fig. 6.1d) are usually favoured. These have a smaller diameter than grabs (less than 15 cm) and consist of an open-ended perspex tube. The top of the tube has a valve which allows the free passage of water during the descent of the corer but which is closed, either automatically or with a messenger, when the sampler is lifted to prevent the loss of sediment. Multiple units, which take several samples at once, have been developed (e.g. Hamilton *et al*, 1970).

Frequently, survey work is carried out in very isolated places and the transport of apparatus becomes a problem. Maitland and Morris (1978) have described a low cost, multipurpose limnological sampler, which can sample water, sediment, phytoplankton, macrophytes, zooplankton, zoobenthos and small fishes. Many of the modules are made of light-weight plastic.

Artificial substrata

Over the last decade considerable interest has been shown in the use of artificial substrata for the collection of macro-invertebrates. A variety of substrata have been tried, including wire mesh trays filled with stones and placed on the river bed (e.g. Hughes, 1975), or wire mesh baskets filled with crushed limestone (Fig. 6.2a), or bark chippings from coniferous trees, placed on the river bed or suspended at various depths in the water (e.g. Mason *et al*, 1967; Bergerson and Galat, 1975). Artificial substrata may be mimics of plants, consisting

(a)

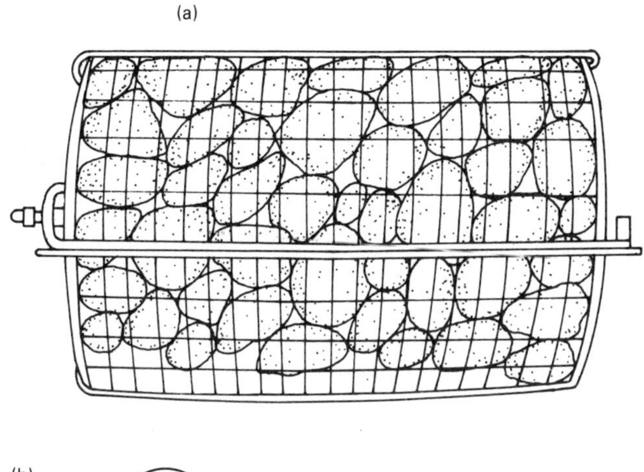

(b)

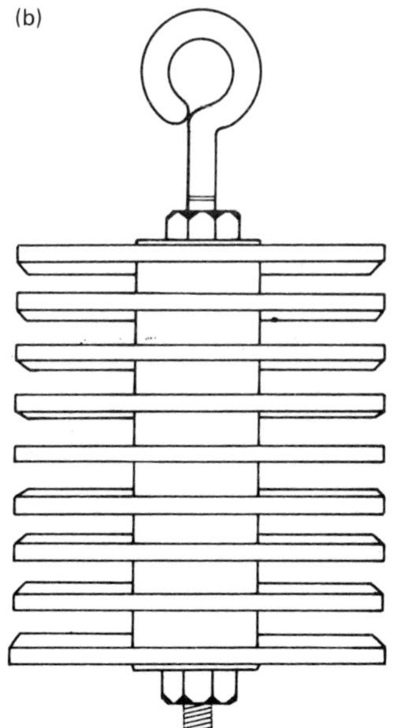

Fig. 6.2. Artificial substrata for sampling invertebrates. (a) rock-filled basket (b) Hester-Dendy multiplate sampler.

of nylon ropes attached to a plastic or wire base (Macan and Kitching, 1972; Mason, 1978). Units of multiple hardboard plates (Fig. 6.2b), suspended in the water, have also been frequently used (Hester and Dendy, 1962). Hellawell (1978) has suggested that standard air-bricks, which have large external surfaces and interstices, make cheap and effective artificial substrata.

Artificial substrata, of course, are initially devoid of invertebrates, so that it is important for a representative invertebrate community to have developed before they are retrieved. Using modified Hester–Dendy multiplate samplers exposed for 60 days in a stream, Meier *et al* (1979) found that the number of individuals present peaked at 39 days and then declined, whereas new taxa were still colonizing at the end of the exposure period. Using a similar collector in a canal, Cover and Harrel (1978) found that the number of species present increased very little after three weeks, while Mason (1978), using artificial macrophytes in a shallow lake, recorded an equilibrium community after 35 days. In these cases, however, species disappeared and new species were added in substrata exposed for long periods while overall populations increased, affecting diversity. It is generally agreed that an exposure period of six weeks will allow a reasonably representative community to develop.

There are a number of advantages in using artificial substrata for the assessment of water quality. The samples can be readily processed because there is usually little extraneous material, such as silt. The effect of natural substrata is reduced and the samples can be used in sites where more conventional techniques are difficult to apply, for instance in deep, swiftly flowing rivers. A higher level of precision is obtained and comparison between sites is made easier. Disadvantages are that the sample obtained may not be representative of the community present in the natural substratum of that site, for example the fauna of stony riffles may colonize artificial substrata positioned in mud and substrata may be selective. They only collect invertebrates colonizing during the period of exposure and this makes comparisons with other techniques difficult. They are subject to vandalism and to loss during spate conditions.

Drift and emergence samplers

Drift and emergence samplers are reviewed by Hellawell (1978). Emergence traps collect only a proportion of the community (i.e. insects) at certain times of the year, while the emergence periods for different species varies, so the technique is of little use in the assessment of water quality. The collection of drifting invertebrates is subject to similar problems in that different species have different predelictions to drift, and temporal, diel, and other factors make the relating of drift to the macro-invertebrate community in a river bed difficult. However, one component of drift, the exuviae (skins) left behind by hatching chironomid pupae, has recently been shown to have considerable potential in the assessment of water quality, as described in Chapter 2. Wilson and McGill (1979) outline the techniques involved, exuviae being collected with a pond net from streams or lakes, especially where concentrations of scum or flotsam occur. As surface samples are taken, the type of water body does not affect collection. The exuviae are easily separated from detritus by sieving. Identification requires considerable expertise, but should become easier when identification keys are made available. Wilson and McGill (1979) showed that the technique can provide a rapid biological assessment of an entire catchment from a single set of field collections taken at a good time of year (late May to September) on a single day. The method also reveals points of change in the fauna due to changes in water quality. It is most effective during the summer and will have little value for the examination of sporadic pollution incidents occurring at other times of year.

A comparison of sampling methods

Very few studies have been made of the efficiencies and relative merits of various samplers. Hughes (1975) compared a Surber sampler, a modified Neil sampler, which was also used in conjunction with an electric shock pulser, and an artificial substratum. The Surber and Neil samplers gave similar results, the electric shock method was highly selective, particularly for mayflies, and was the least consistent. The artificial substratum collected the most species and numbers and was the most consistent, but Hughes considered that it failed to represent the fauna in the surrounding river bed and so preferred the Surber and Neil samplers.

Flanagan (1970) compared a range of core samplers and grabs for sampling sediments beneath deep waters, core samples also being taken by a diver. The Ekman grab and multiple corer gave good quantitative estimates of total biomass when compared with the diver. The Ekman grab collected fewer oligochaetes and the corer fewer chironomids. Overall the multiple corer appeared to have better performance and was the preferred sampler in Hellawell's (1978) review, a conclusion also reached by Maitland (1978).

Several studies have compared different types of artificial substrata (e.g. Prins and Black, 1971; Beak *et al*, 1973; Mason *et al*, 1973; Crossman and Cairns, 1974). In general, all samplers that have been tested have developed a diverse fauna, but marked differences between the collections of various substrata at the same site may occur. The sample variability between artificial substrata at a site is relatively low compared with samples collected by traditional methods so that a small number of replicates (e.g. three in the case of Mason *et al*, 1973) will give a representative sample of the macrofauna. If the aim of a study is to assess water quality, rather than examine the local macrofauna, then artificial substrata have much to recommend them because, even if the sample collection is richer than in the surrounding river bed, as it frequently will be if artificial substrata are placed on depositing beds, it shows that a diverse fauna is capable of living in those water quality conditions. As Green (1979) has succinctly remarked, 'the health of canaries can be a good indicator of the safety of coal mines, even though canaries are not natural inhabitants of coal mines'. For extensive comparative surveys, however, it will be necessary to standardize the artificial substrata and their placement within a water body, but this standardization is easier than with traditional sampling methods.

Sorting samples

The sorting of samples is a tedious and time-consuming occupation, especially of material from depositing substrata. Coarse detritus can be removed using a 4 mm sieve, but a smaller sieve to retain invertebrates has to be a compromise between retention efficiency and the time available for sorting, bearing in mind that hand-sorting will be inefficient if large amounts of detritus remain with the sample. Jónasson (1955) found the retention efficiency of chironomids to be doubled if the mesh opening of the sieve was reduced from 620 μm to 510 μm. Mason (1977a) found that 87 per cent of chironomids and 82 per cent of tubificid worms were retained by a 500 μm aperture sieve and, using a 250μm sieve, the retention efficiency was 98 per cent and 99 per cent respectively. The smaller mesh sieves readily become clogged with clay particles so that wet sieving becomes very time consuming. For routine surveillance programmes a 500 μm aperture sieve should be adequate.

To further separate invertebrates from the substratum a variety of flotation techniques have been tried, using solutions such as calcium chloride, carbon tetrachloride or sucrose (Hellawell, 1978), but these are likely to add another source of loss to the sorting programme and can, in themselves, be time-consuming. The most effective method is probably to transfer the contents of a 500 μm sieve, using a gentle jet of water, to a white tray, which has been divided into squares, and then to sort each square systematically under a good light. A stain, such as rose bengal, can be added to help sorting, but where possible the sorting of a live sample is best because the movement of animals aids detection.

Data processing

Surveillance programmes generate large volumes of data, which are best stored and analysed using computers. Archiving systems for storing water quality data have been developed in the USA and Great Britain and the advantages are discussed by Pearson (1978). Coded checklists, which aid in storing biological data in a uniform way, have been provided in Great Britain on freshwater animals by Maitland (1977), on algae by Whitton *et al* (1978) and on freshwater macrophytes by Holmes *et al* (1978).

The analysis of surveillance data may be done by multivariate techniques or by using biotic or diversity indices, the latter two methods being most widely used in the water industry. Biologists in the water industry, as already stated, must communicate their findings to managers and the public, who are usually unfamiliar with ecological techniques. A single figure describing the biological impact of water quality at a site is therefore a tempting way to present data and is generally preferred by water managers (Thomas, 1976). It does, however, reduce the amount of information extracted from the data, and the ready calculation of indices tends to replace more sophisticated analyses. These would enable the relationships between organisms and the measured physico-chemical parameters of water quality to be better understood, eventually placing the biological management of freshwaters on a sounder footing. Green (1979), who is decidedly unenthusiastic about derived indices, has stated, 'the literature has been filled with wrangling over which diversity index should be used of the countless ones proposed while there has been too little emphasis placed on the critical importance of proper sampling design, or on appropriate multivariate statistical analysis methods that can efficiently test exactly the hypotheses one wishes to test in environmental studies'. Biotic and diversity indices can be used for overall assessments of water quality and their ability to detect changes in quality will be illustrated later in this chapter, but it must be stressed that information, and indeed the expertise of biologists, is tragically wasted if indices become the be-all and end-all of biological surveillance programmes, as they so often are in the water industry.

Multivariate analyses

Multivariate analyses can be carried out on both presence-absence and quantitative data and a number of authors (e.g. Hurlbert, 1969) suggest that, as abundance is easily influenced by extraneous factors, presence-absence data give a less ambiguous measure of association. Multivariate techniques can identify discontinuities present within communities which can be related to environmental change. They can be used to generate hypotheses about the causality of distribution, but the relationship of distribution to environmental features must then be studied using experimental techniques.

Green (1979) strongly recommended that principal components analysis (PCA) should form the basis of a multivariate analysis, the principal component score being the variable which can be subjected to ordination, clustering or other statistical techniques. PCA summarizes sets of correlations between variates. A principal component is an additive combination of $a_1x_1 + a_2x_2 + \cdots a_nx_n$ of the n original variables, with the coefficients ($a_1, a_2, \ldots a_n$) chosen so that, as nearly as possible, a_ia_j equals the correlation r_ir_j between the ith and jth variables. The first principal component is chosen to make the agreement as close as possible and so on until n principal components have been calculated for the n variables. Together they reproduce the observed set of correlations which are usually summarized closely by the first two or three principal components. The method is explained by Williamson (1972).

Coefficients of similarity can be calculated before a cluster analysis is carried out. Kaesler *et al* (1971), Kaesler and Cairns (1972), Hocutt *et al* (1974) and Edwards *et al* (1975) have used the Jaccard coefficient, S_j (Jaccard, 1908) for comparing community species lists:

$$S_j = \frac{a}{a + b - c} \qquad [6.3]$$

where a is the number of species in community A, b is the number of species in community B and c is the number of species common to both communities.

The Sørensen 'quotient of similarity', QS (Sørensen, 1948) has been used by Bryce *et al* (1978), where:

$$QS = \frac{2c}{a + b} \qquad [6.4]$$

Mason and Bryant (1974), Fahy (1975) and Jones and Peters (1977) have used the index of affinity of Fager (1957),

$$I = \frac{2J}{nA + nB} \qquad [6.5]$$

where J is the number of joint occurrences of species A and species B, and nA and nB are the total number of occurrences of species A and species B respectively.

Using a clustering technique based on Jaccard coefficients for macro-invertebrates collected from the River Cynon in South Wales, Edwards *et al* (1975) were able to identify three primary groups (Fig. 6.3), the species being listed in Table 6.1. Group C was abundant in the upper stretches of the river, but declined downstream, probably due to the intermittent release of coal washings. Group A increased downstream, associated with other pollutional factors. The study of this river, receiving a complex mixture of pollutants, is described in more detail in Chapter 2.

Bryce *et al* (1978) used a single linkage clustering technique of quotients of similarity to examine the macro-invertebrate fauna of the River Lee in south-east England and a dendrogram of their results is shown in Fig. 6.4. Stations 2–8 and station 10, on a tributary, formed a faunistically close group in the upper part of the river, where the water was free of pollution. A second grouping was formed by stations 9, 11–15 and 17, which were in a clean-water, but canalized, section of the lower river. Station 16, on a tributary with a silted substratum, also fell into this group. Stations 18–20 made up a third group occurring in a polluted stretch of the river. Station 1 was isolated from the other sites and received a considerable quantity of urban run-off, but no pollution by sewage effluent.

The association of invertebrate groups with the flow regimes of rivers (Jones and Peters, 1977) has already been described (p. 62). Multivariate techniques to examine the relationship between associations of species and water quality parameters have rarely been used

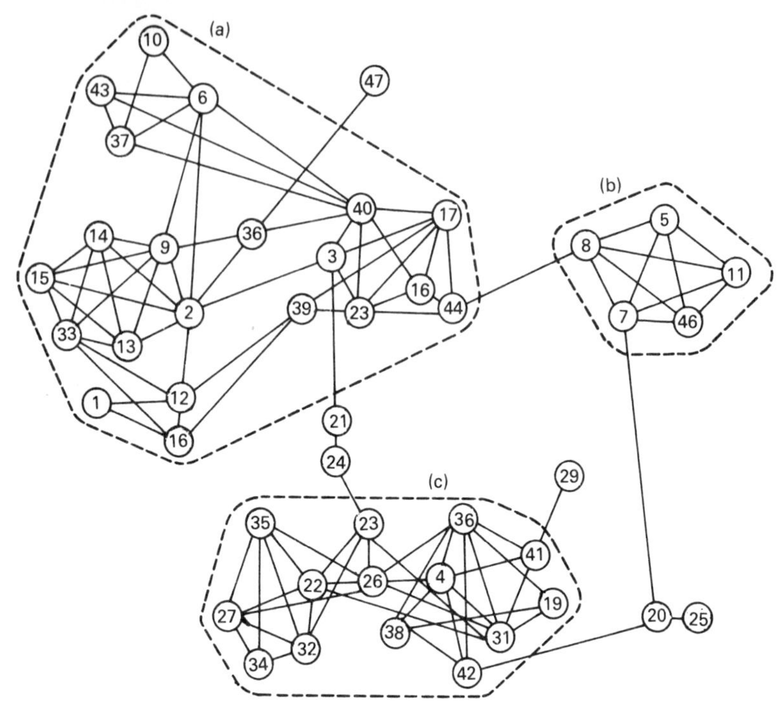

Fig. 6.3. Associations of macro-invertebrates in the River Cynon, South Wales (from Edwards *et al*, 1975).

but they offer considerable scope for aiding the understanding of pollutional phenomena. General introductions to multivariate statistics are those of Anderson (1960) and Clifford and Stephenson (1975), while Goldsmith and Harrison (1976) provide a review of their use in the analysis of vegetation, a field in which multivariate techniques have been widely applied.

Biotic indices

A number of biotic indices have been developed for the assessment of water quality and they have been reviewed by Hellawell (1978). A biotic index takes account of the sensitivity or tolerance of individual species or groups to pollution and assigns them a value, the sum of which gives an index of pollution for a site. The data may be qualitative (presence-absence) or quantitative (relative abundance or absolute density). They have been designed mainly to assess organic pollution.

Table 6.1. The macro-invertebrate associations in the River Cynon, South Wales (from Edwards *et al.* 1975).

A	*B*	*C*
Hydra sp.	*Nais alpina*	*Chaetogaster langi*
Chaetogaster crystallinus	*Nais communis*	*Leuctra fusca*
Chaetogaster diaphanus	*Nais elinguis*	*Ephemerella ignita*
Nais barbata	*Pristina idrensis*	*Ecdyonurus dispar*
Nais variabilis	*Potamopyrgus jenkinsi*	*Oreodytes septentrionalis*
Pristina foreli		*Limnius volckmari*
Stylaria lacustris		*Trissopelopia longimana*
Limnodrilus hoffmeisteri		*Polypedilum acutum*
Tubifex tubifex		*Brillia modesta*
Lumbricillus rivalis		*Eukiefferiella calvescens*
Helobdella stagnalis		*Orthocladius* sp.A.
Acanthocyclops vernalis		*Bezzia-Palpomyia* gp.
Eucyclops speratus		*Wiedemannia* sp.
Prodiamesa olivacea		*Ancylus fluviatilis*
Brillia longifurca		
Cricotopus bicinctus		
Orthocladius rubicundus		
Rheocricotopus foveatus		
Syncricotopus rufiventris		
Hygrobates fluviatilis		
Levertia porosa		

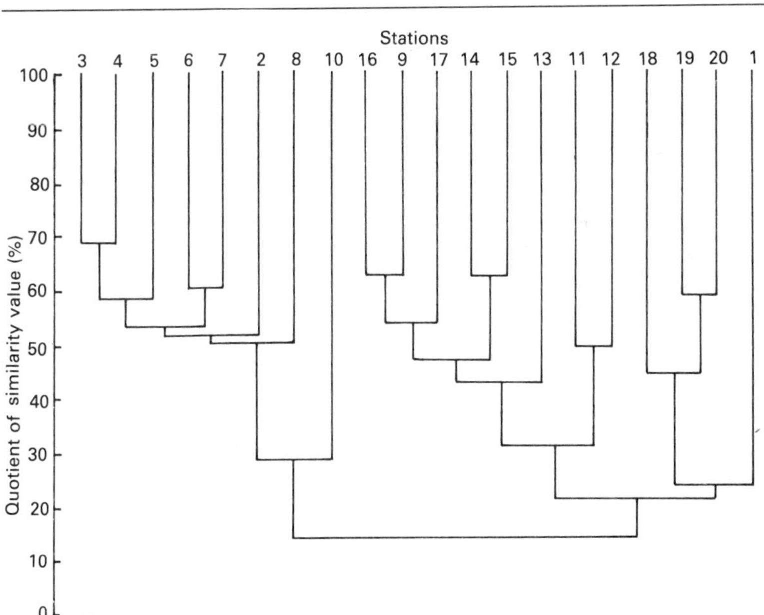

Fig. 6.4 Dendrogram based on values of quotient of similarity for all species of macro-invertebrates recorded at 20 stations along the River Lee, south-east England (from Bryce *et al*, 1978).

The earliest devised biotic index was the Saprobien system of Kolkwitz and Marsson (1908, 1909), who recognized four stages in the oxidation of organic matter—polysaprobic, α-mesosaprobic, β-mesosaprobic and oligosaprobic. The presence or absence of indicator species in the zones is recorded. Pantle and Buck (1955) developed the Saprobien system to take into account the relative abundance of organisms in a sample. They gave a value (h) to express the abundance of each organism in the different Saprobien groups as well as a value (s) for the saprobic grouping (Table 6.2). The Saprobien index and derivatives are widely used in continental Europe but the method has received little support in Britain or North America. Saprobien methods are reviewed by Sládeček (1979).

In Britain, the two most widely used biotic indices are the Trent Biotic Index (TBI) and the Chandler Biotic Score (CBS). The Trent Biotic Index was developed by Woodiwiss (1964) for use in the Trent river system in midland England but it is now applied to surveillance data from rivers throughout the United Kingdom. It is relatively simple to apply and takes into account the presence and absence of species and species richness, but animals do not need counting. The sensitivity to organic pollution of different species or groups is used in determining the index. In grossly polluted waters, where no macro-invertebrates are present, a TBI of zero is obtained. The maximum score, in unpolluted water with a species rich invertebrate fauna, is 10. As no account is taken of abundance, a single individual of a sensitive species (e.g. a stonefly larva drifting downstream) can have a disproportionate effect on the index. A worked example of the TBI is given in Appendix A.

The abundance of organisms within the community, as well as the species richness, is of value in assessing the degree of pollution.

Table 6.2. The saprobic index of Pantle and Buck (1955).

	s value		h value
Oligosaprobic	1	Occurring incidentally	1
β-mesosaprobic	2	Occurring frequently	3
α-mesosaprobic	3	Occurring abundantly	5
Polysaprobic	4		

$$\text{Mean saprobic index } (S) = \frac{sh}{h}$$

The saprobic index ranges are:

1.0–1.5	Oligosaprobic	no pollution
1.5–2.5	β-mesosaprobic	weak organic pollution
2.5–3.5	α-mesosaprobic	strong organic pollution
3.5–4	Polysaprobic	very strong organic pollution

Chandler (1970) produced an index which has five levels of abundance, the score of each indicator species being weighted in relation to its abundance. A worked example is given in Appendix B. If a species, intolerant of pollution (e.g. a stonefly), is abundant it is given a high score (100), whereas an abundant, pollution tolerant species (e.g. *Chironomus riparius*) is given a low score (4). The allocation of scores is somewhat arbitrary. As taxa have to be counted as well as identified, the score takes much longer to determine than the TBI. The lower limit of the score is zero, when no macro-invertebrates are present, while there is no upper limit. Several modifications of the Chandler Biotic Score have been made. The score can be averaged by dividing the total score by the number of taxa, the maximum value then being 100 (Cook, 1976). Bryce *et al* (1978) have reorganized the scores into taxonomic groupings, have compressed the number of abundance classes to three and have revised some of the scores. Hargreaves *et al* (1979) simplified the score for use in the field. Thirteen groups of organisms were arranged in order of increasing tolerance to organic pollution and scores corresponding to four different abundance classes were assigned to each group. Each group was considered as a single entity for the purposes of the index so that the abundance of constituent organisms did not need to be separately determined. Relative abundance as a percentage was used to minimize the effects of variations in sample size. A maximum score of 645 is attainable. A worked example is given in Appendix C.

Diversity indices

Biotic indices have been developed largely to measure responses to organic pollution and may be unsuitable for detecting other forms of pollution. For example, stoneflies play a prominent role in most biotic indices because of their great sensitivity to a decrease in oxygen in the water, but they are markedly more tolerant of metals and may be abundant in rivers receiving substantial quantities of metal wastes from mine tips. Diversity indices are used to measure stress in the environment. It is considered that unpolluted environments are characterized by a large number of species, with no single species making up the majority of the community and a maximum diversity is obtained when a large number of species occur in relatively low numbers in a community. When an environment becomes stressed, species sensitive to that particular stress will be eliminated, thus reducing the richness of the community and certain species may be favoured (e.g. with the reduction of competition or predation) so that they become abundant compared with other members of the community. The majority of species diversity indices take account of both the number of species in a sample and their relative abundances, but the sensitivity of individual species to particular pollutants is not allowed for. Diversity indices are widely used by North American workers and they are frequently calculated, along with biotic indices, by British workers.

The simplest index of diversity is the sequential comparison index (SCI) of Cairns *et al* (1968), which is used to estimate relative differences in diversity between sites and requires no taxonomic knowledge. The observer distinguishes between consecutive individuals in the sample on the basis of shape, size and colour. If the currently observed individual is identical to the previous individual it forms part of the same run, if different the current animal forms the beginning of a new run. The sequential comparison index is:

$$\text{SCI} = \frac{\text{number of runs}}{\text{number of individuals}}$$

The method is detailed in Cairns and Dickson (1971). It is aimed at the non-biologist and there is no taxonomic input so that closely related species, which differ in their sensitivity to pollution, would not be detected. The approach is very unsubtle. Resh and Unzicker (1975) stress the importance of species identification in water pollution studies.

The most widely used indices of diversity are those based on information theory and were introduced into water quality assessments by Wilhm and Dorris (1968). The most frequently used measure is:

$$H' = -\sum_{i=1}^{s} pi \log pi \qquad [6.6]$$

where pi is estimated from ni/N as the proportion of the total population of N individuals belonging to the ith species (n_i). Wilhm and Dorris (1968), after examining diversity in a range of polluted and unpolluted streams, concluded that a value of H' greater than 3 indicated clean water, values in the range 1–3 were characteristic of moderately polluted conditions and values of less than 1 characterized heavily polluted conditions. Different base logs are used in diversity indices so care is needed in comparisons. A worked example is given in Appendix D.

The abundance of organisms is obviously important in assessing the effects of pollution, but it can make the interpretation of diversity indices very difficult, especially when water quality over time at a particular site is being examined or when diversity indices from a wider geographical area are being compared from samples taken at different times of year. There may be marked changes in the seasonal abundance of animals. Over two years, Mason (1977b) examined the diversity of monthly samples of macro-invertebrates collected from a hypereutropic lake, devoid of submerged plants, and a clear-water, eutrophic lake with rich macrophytic growth. Diversity was generally lower at the hypereutrophic site, but in June of both years the diversity index was lower at the unpolluted site, due to the presence of a very high population of the chironomid larva *Tanytarsus holochlorus*, which developed rapidly and then emerged from the lake. When sampling is infrequent, as with most surveillance programmes, the appearance of seasonally abundant species could result in the misinterpretation of water quality conditions using diversity indices. Hughes (1978) found that sampling method, the area sampled, the time of year and the level of identification all influenced the diversity index, while Murphy (1978) showed that the seasonal variations in the index at a site were greater than differences between sites along a river. Extreme care is obviously needed in the interpretation of diversity indices. Mason (1977b) concluded that the number of species (S) alone gave a more consistent indication of the difference in eutrophic status of two lakes and Winner *et al* (1975) drew the same conclusions in a study of streams polluted with copper. Green (1979) quoted a number of studies which suggest that S is a better and more realistic indicator of diversity than information statistics.

A comparison of indices

Several studies have compared indices in an attempt to determine which method might best reflect water quality over a wide range of conditions. Nuttall and Purves (1974), in a survey of the macro-invertebrates of the Tamar catchment compared the Trent Biotic Index, Chandler Biotic Score and a diversity index and considered that the last of these was the least likely to produce anomalous results, the other two indices failing to pick up point sources of mild pollution. Samples were taken at only two times of year. Murphy (1978) sampled two rivers at four times of year and compared three diversity indices, the Trent Biotic Index, the Chandler Biotic Score and an averaged Chandler Biotic Score. The diversity indices showed such temporal variability that spatial variation was masked. The TBI was found to be insensitive, but the CBS and averaged CBS showed little temporal variability and reflected a downstream deterioration in the quality of water. Murphy recommended the use of the averaged CBS which responded to changes in water quality, and, unlike the CBS, was largely independent of the number of species collected.

Cook (1976) used the CBS, which was designed for use in southern Scotland, and an averaged CBS, to examine a stream in New York State, USA, where the species were different, though of similar taxonomic types. The scores were compared with a diversity index, using monthly samples collected from eight stations over a year, and Cook concluded that the averaged CBS was the most sensitive to variables influenced by pollution.

Balloch *et al* (1976) assessed several indices, including TBI, CBS and a diversity index, on rivers in England, Scotland and Wales. The results for the North Esk, a Scottish river receiving a number of polluting discharges, are shown in Fig. 6.5. The TBI did not reflect the gradual improvement of the mildly polluted conditions in I, J and

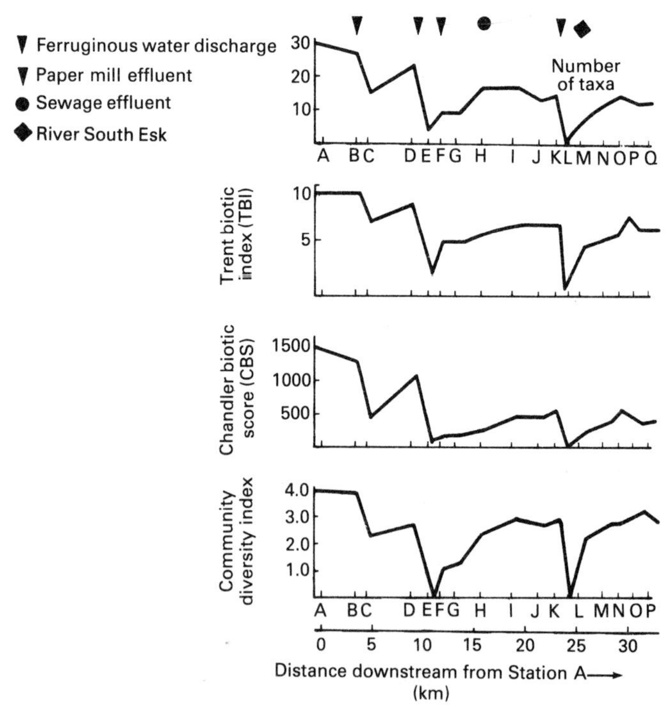

Fig. 6.5. A comparison of three biological indices at 17 stations down the River North Esk, Scotland (adapted from Balloch *et al*, 1976).

K, whereas these were picked up in a gradual rise in the CBS. On the mildly polluted English and Welsh rivers the TBI was also insensitive to slight changes in quality. The diversity index reflected the changes in water quality in the North Esk, but indicated almost complete recovery by station J (with a similar index to station D, above the effluent outfall), though this was certainly not the case. Balloch *et al* concluded that the CBS was the most sensitive index to changes in water quality associated with mild and moderate conditions of pollution.

In general, critical comparisons of biotic indices indicate that the CBS, or its modifications, is the best generalized index for the routine assessment of water quality. The reader might, however, like to compare the top graph (number of taxa, *S*) in Fig. 6.5 with the three indices and decide for himself whether they convey more information.

Conclusions

Only a few of the multiplicity of approaches to processing data in routine surveillance programmes have been presented here and new developments and modifications of old methods are continually being tried. Despite their very widespread use, I remain unconvinced that biotic and diversity indices tell us much more about the biological aspects of water quality than the number of taxa (*S*) alone and *S* assumes nothing concerning the tolerance of species to particular pollutants, the temporal changes in population size, or the theoretical organization of communities.

Having examined several water quality indices Ghetti and Bonazzi (1977) concluded that 'in the comparison between the various criteria for reading biological data, it can be seen that, once again, in this field every conceptual or methodological simplification, for example in order to obtain applicative results, entails a considerable loss of information and a wide margin of discrepancy in the ensuing results. For this reason, it is mistaken to pretend that these practical methods can produce answers whose accuracy can be gauged at will'.

The number of species, preferably with estimates of abundance, can be used to examine the relationships between the biological community and the water quality characteristics of sites using multivariate techniques to generate hypotheses which can then be tested experimentally. It is only in this way that a detailed understanding of the effects of pollution on natural communities can be developed and used in the rational management of freshwaters.

Chapter 7

TECHNIQUES FOR INTRUSIVE MICRO-ORGANISMS AND BIO-ASSAYS

Introduction

The previous chapter was concerned mainly with methods of surveillance in the field and considered especially the sampling and enumeration of invertebrates, which appear on balance to be the most practical group to use. The present chapter will consider those methods in which most of the effort is in the laboratory, i.e., intrusive micro-organisms, and bio-assay tests, and those methods involving the introduction of test organisms into the environment.

Intrusive micro-organisms

Bacteria

Despite the ubiquity of bacteria in aquatic ecosystems and the large populations developed, little attention has been given to the use of indigenous bacteria in the assessment of pollution. The obvious exception is in the determination of the biochemical oxygen demand (BOD), which is used to determine the pollutional potential of a waste water containing an available source of organic carbon, by measuring the amount of oxygen utilized by micro-organisms in a standard sample. BOD provides a broad measure of the effects of organic pollution on a receiving water. The methods for determining BOD are given in Department of the Environment (1973) and American Public Health Association *et al* (1976), while the rationale, development and limitations of the test are discussed by Gaudy (1972) and Stones (1979).

Recent work has suggested that bacteria which metabolize petroleum hydrocarbons are ubiquitous in aquatic ecosystems but are present in much higher numbers in environments exposed to oil (Buckley *et al*, 1976; Bartha and Atlas, 1977). The development of techniques using the responses of indigenous bacteria to pollution could prove invaluable for pollution evaluation.

By contrast, the use of intrusive bacteria in the assessment of faecal pollution of freshwaters has received very wide attention. A number of bacteria, viruses and invertebrate parasites present a potential public health hazard (p. 17) and it is essential that water, if it is used by the public (e.g. for drinking) should be free of pathogens. Pathogens may occur only in very low numbers, or intermittently, so other microbiological indicators of faecal pollution are used. To indicate the potential occurrence of pathogens, indicators must satisfy certain criteria (Bonde, 1977):

1. they should always be present when pathogens are present and occur in greater numbers than pathogens;
2. they should not proliferate to any greater extent than pathogens in the aquatic environment;
3. they must be more resistant to aqueous environments and disinfectants than pathogens;
4. they should give simple and characteristic reactions, enabling, as far as possible, unambiguous identification, and they should grow readily on relatively simple media;
5. they should grow independently of other organisms present on artificial media.

The most frequently used indicators of faecal pollution are the coliform bacteria, faecal streptococci and *Clostridium perfringens*.

Coliform bacteria are Gram-negative, oxidase-negative, non-sporing rods, which are able to ferment lactose at 35–37 °C in 48 h, with the production of acid and gas. In the past the total coliform count has been used as an indicator of faecal pollution, but Dutka (1973) has shown that coliforms failed to satisfy the criteria for indicators of potential pathogens given above, because several types of coliform bacteria are non-faecal in origin and pathogens have been found in water when coliforms were absent. Some coliforms can also multiply in natural waters.

The presence in water of faecal coliforms, especially *Escherichia coli*, is a better indicator of sewage contamination than total coliforms (Bonde, 1977). Human faeces always contain high numbers of *E. coli* and the organisms can be readily distinguished from other coliforms. The relationship between *E. coli* counts and numbers of pathogenic *Salmonella* from one set of data is shown in Fig. 7.1.

Faecal streptococci are also used as indicators of faecal pollution and it is possible to distinguish between streptococci from human and

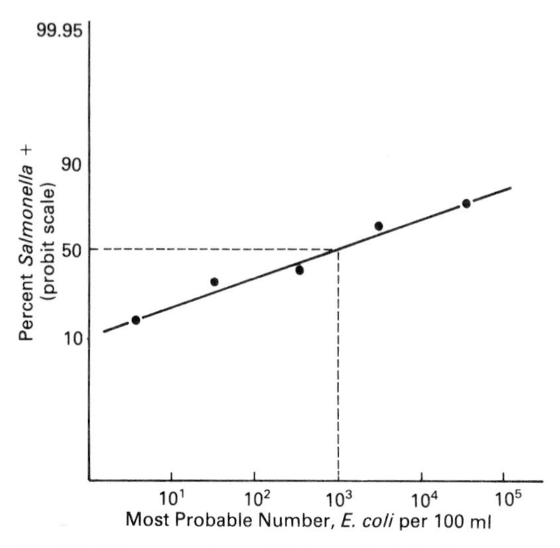

Fig. 7.1. The percentage of *Salmonella* isolations versus counts of faecal coliforms (from Bonde, 1977).

animal sources, so that the origins of pollution can frequently be determined. Streptococci die off rapidly outside the intestine so their presence indicates recent contamination of a watercourse. The ratio of faecal coliforms to faecal streptococci has been used to determine the origin of faecal pollution. An FC:FS ratio greater than 4 may indicate a discharge of municipal wastes, while an FC:FS ratio of less than 0.6 may be indicative of storm water run-off (Geldreich and Kenner, 1969).

The anaerobic sporeformer *Clostridium perfringens* has been suggested as an alternative or accompaniment to *E. coli*, particularly where remote or old pollution is being examined. Bonde (1977) considered that *Cl. perfringens* is a better indicator than *E. coli* when toxic substances are present in the water and if the transport of samples to the laboratory takes more than 12 hours.

A number of other bacteria, e.g. *Bifidobacterium* sp., *Pseudomonas aeruginosa*, *Salmonella* sp. have been considered as potential indicators, but none of them fully satisfy the criteria listed above.

Bacterial populations can be estimated by directly counting all the visible bacteria in a counting chamber or in a fixed, stained preparation, giving the total count; or by determining the number of living bacteria, the viable count. Two methods are available for viable counting, either by counting the colonies which grow when a solid medium is inoculated with a known volume, or by incubating a series of dilutions in a fluid medium and observing the appearance or non-appearance of growth, the true bacterial count then being estimated by the most probable numbers MPN) technique. Bonde (1977) and Jones (1979) describe these methods, examine their statistical basis and recommend when particular techniques should be used. Bonde also describes the media and methods for particular indicator species, information which is also contained in Department of the Environment (1973) and American Public Health Association (1976).

Most work on bacterial indicators has involved an examination of the water but it appears that sediments accumulate bacteria, so that the numbers of bacteria in the sediment are often greater than in the overlying water (Van Donsel and Geldreich, 1971; Gerba *et al*, 1977; Matson *et al*, 1978). Thus counts of indicators from sediments may give an average value of the numbers of bacteria continually being sedimented and may be less variable than counts from the water, though there are difficulties in enumerating bacteria from sediment samples.

Viruses

More than 100 types of enteric viruses are excreted in human faeces (Bitton, 1978) and they represent a considerable potential health hazard in water used for potable supply and recreation, especially as viruses survive the sewage treatment process well. Viruses are present in water in much lower numbers than bacteria and large volumes of water must be examined to detect them, but low-level transmission of viruses by drinking water may be occurring at a much greater frequency than was previously thought (Metcalf, 1978) and lower levels of virus contamination than bacteria contamination may cause infections.

Methods for the detection of viruses in water have been reviewed by Hill *et al* (1971) and Shuval and Katzenelson (1972). In most cases, the presence of faecal coliforms is taken to indicate that viruses of human origin are also likely to be present in water, but there are dangers in this approach for viruses may survive for much longer in water than intrusive bacteria. Shuval (1975) has calculated that bathing in waters containing 10 000 coliforms/100 ml might give viral gastro-enteritis in one per thousand bathers on a particular day and, with lower coliform counts, the incidence of viral infection is correspondingly reduced. The methodology for detecting viruses is not yet suitable for routine laboratory tests and methods are both time consuming and expensive.

Bio-accumulators

The concentration of poisons in the tissues of organisms is discussed on p. 120–126. Where pollutant levels in water are near the limit of detection, the analysis of tissues can bring detection within the scope of most instruments and operators and this is particularly advantageous for surveillance programmes involving a number of laboratories with widely differing facilities and expertise. Whereas the analysis of a water sample will record the level of a pollutant at a particular time, the bio-accumulator will reflect the level of pollution over much longer periods. Furthermore, dried tissues can be stored for long periods before analysis, whereas water samples need almost immediate analysis, particularly as transformations of material due to microbial activity may occur. Organisms are often monitored to assess the risk to the public of exposure to material with which they are likely to come into contact, *critical material*, such as food.

A number of criteria need to be fulfilled before an organism can be considered a satisfactory biomonitor. Organisms should be relatively sedentary so that they reflect only local pollutant levels. They need to be readily identifiable and in sufficiently large numbers to ensure genetic stability. They also need to be sufficiently large so that low concentrations of pollutants can be detected within individuals. The life cycles should be long enough to ensure that there is a good balance of age groups throughout the population during the monitoring period. The degree with which organisms concentrate pollutants will vary with both the pollutant and the species so, for large-scale surveys, a single, widespread species is needed. Although the use of bioaccumulators has considerable appeal, the results are often difficult to interpret. The total pollutant content and concentration may vary with the age of an organism, its size, weight and sex as well as the time of year, the sampling position and the relative level of other pollutants in the tissues.

In addition to measuring pollutant levels in organisms *in situ*, they can also be placed into the environment in cages and uptake rates can be measured over defined periods of time. This approach can be very valuable in comparing pollution levels between sites and at different times of year.

Most effort in the use of bio-accumulators in monitoring programmes has thus far been in the coastal marine environment (e.g., reviews by Phillips, 1977; Wright, 1978; Perkins, 1979), because of the economically important fisheries and shellfisheries in these regions. Intertidal macrophytic algae, such as *Fucus*, and bivalves, especially *Mytilus*, have received particular study because they satisfy the criteria given above.

Work in the freshwater field has also concentrated on macrophytes and molluscs as bio-monitors. Ray and White (1979) have suggested the pteridophyte *Equisetum arvense* as a valuable monitor for heavy metals. This horsetail grows along watersides and puts out large rhizomes, which are in contact with the water-saturated substratum. Ray and White considered that differences in metal content in plants growing by a river system receiving acid mine drainage reflected the integrated metal concentrations in the water over a long period. Wolverton and McDonald (1978) used the water hyacinth, *Eichhornia crassipes*, a plant growing within the water, as a bio-magnifier of cadmium, the root system of the plant concentrating the metal very quickly. At low concentrations of 0.001 mg l^{-1} cadmium chloride, the plant's root system concentrated the element at an average rate of 0.9, 1.4 and 3.0 μg Cd g^{-1} dry weight after 24, 48 and 72 h exposure respectively, and at higher environmental levels the rate of concentration was very much faster. Särkka (1979) considered the freshwater mussel *Anodonta piscinalis* and the sponge *Spongilla lacustris* to be excellent species for monitoring chlorinated hydrocarbon pesticides because of their bio-accumulation rates, sedentary behaviour and ease of sampling by scuba diving, while for short-term bio-accumulation studies of cadmium, Poldoski (1979) recommended the use of the planktonic crustacean *Daphnia magna*.

Foster and Bates (1978) used the freshwater mussel *Quadrula quadrula* for monitoring copper wastes from an electroplating works. They placed animals in cages at several distances below the point of discharge (Fig. 7.2). At 0.1 km below the discharge the mussels accumulated 20.64 μg Cd g^{-1} in 14 days and died. The amount of copper accumulated with time decreased at greater distances from the outfall. At no time did the copper in the effluent exceed the legal limits laid down, so the bio-accumulators were indicating how very low levels of the metal were having adverse effects on the stream fauna. A study of the native mussel fauna showed that the effluent resulted in lethal body concentrations of copper for 21 km below the outfall and, as mussels are very slow to recolonize, the stability of the native fauna had been seriously altered and recovery could take several years.

Although there are many problems associated with the use of bio-accumulators, they are of considerable potential value, especially if caged organisms, standardized for age, condition, etc., are used, as in the study by Foster and Bates.

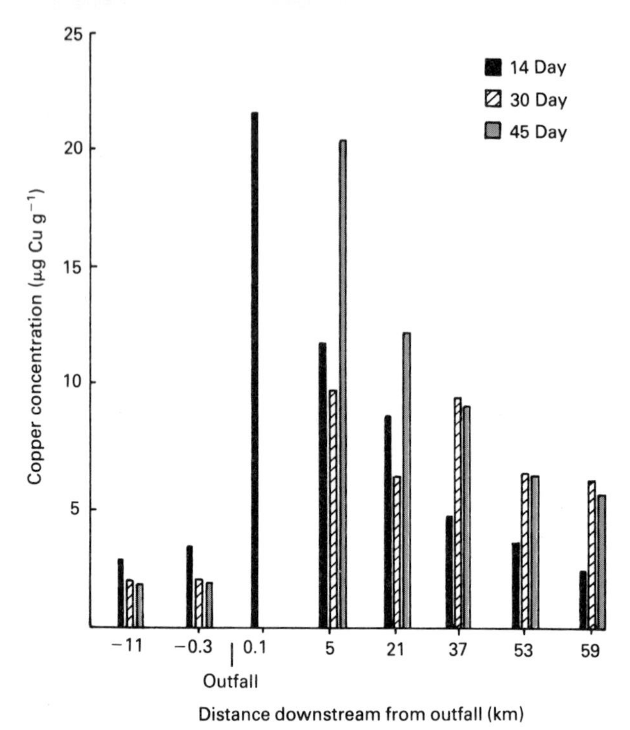

Fig. 7.2. The accumulation of copper in the tissues of caged mussels, *Quadrula quadrula*, exposed in the Muskingum River, USA (from Foster and Bates, 1978).

Bio-assay

Because pollutants often occur in complex mixtures and interact, the chemical determination of pollutant levels alone will frequently give little indication of their potential biological effects. In a bio-assay, environmental conditions are carefully controlled so that the response of a test organism to particular pollutants can be defined, but the extrapolation from bio-assay to field situations may be dubious. Bio-assay techniques must go hand in hand with field observations and field experiments in order to fully understand a pollution problem.

The selection of a suitable organism for routine bio-assays will depend on a number of factors (American Public Health Association, 1976):

1. the organism must be sensitive to the material or environmental factors under consideration;

2. it must be widely distributed and readily available in good numbers throughout the year;
3. it should have economic, recreational or ecological importance both locally and nationally;
4. it should be easily cultured in the laboratory;
5. it should be in good condition, free from parasites or disease;
6. it should be suitable for bio-assay testing.

Small organisms with short generation times are generally preferred for bio-assays, though fish are also popular because of their physiology and their recreational and economic importance. The procedures for selecting, collecting and handling organisms prior to bio-assays are given in American Public Health Association (1976).

Bio-assay tests can be divided into bio-stimulation and toxicity tests.

Bio-stimulation

Bio-stimulation tests are mostly carried out using algae and are suitable for evaluating the nutrient status of a water body, for distinguishing between total and biologically available nutrients and for determining the potential effects of changing water quality on algal growth (Bellinger, 1979). Algal bio-assays can be done in the field (e.g. Lund tubes, p. 85–87), but most involve laboratory studies under defined conditions.

The Standard Bottle Test (United States Environmental Protection Agency, 1971) has been widely used and is a measure of either maximum specific growth rate or maximum standing crop. The specific growth rate (μ) is:

$$\mu = \frac{\ln (x_2/x_1)}{t_2 - t_1} \tag{7.1}$$

where x_1 is the initial biomass concentration at time t_1 and x_2 is the final biomass concentration at time t_2. The maximum standing crop is the maximum algal biomass achieved during incubation. Biomass may be obtained by the direct determination of dry weight, by counting cells, by absorbance, by chlorophyll measurements or by total cell carbon. Standard test algae are normally used, *viz Selenastrum capricornutum, Asterionella flos-aquae, Microcystis aeruginosa*, or *Anabaena flos-aquae*, of which the first is easiest to culture and use (Payne, 1975). To use test algae, the algae initially present in the test water must be removed, by membrane filtration or autoclaving, which may alter the quality of the water. Nutrient depletion and the build up of metabolic wastes also occur during the test period in the closed bottle. To minimize the effects of pretreatment, Lund (1959) recommended removing algae by heating unfiltered water to 40–45 °C for one hour prior to use, while to prevent nutrient depletion and the build-up of wastes, Bellinger (1979) used a dialysis membrane, which allowed nutrients and waste products to diffuse in and out of the test cell, giving a growth rate 46–57 per cent higher than the standard bottle.

Payne (1975) has evaluated the standard bottle test. Figure 7.3 shows the effect of adding primary and secondary treated sewage effluents in water from the oligotrophic Lake Tahoe and the eutrophic Clear Lake in California. Both effluents resulted in a large increase in the populations of *Selenastrum* in Tahoe water, but there was no effect in Clear Lake water. The effect of tertiary treatment was examined in bio-assays by using water from Lawrence Lake, Michigan (Fig. 7.4), where spiking with tertiary treated sewage produced no significant growth, compared with the large growth when the lake water was spiked with secondary treated sewage.

Klekowski (1978) has recently suggested that the fern, *Osmunda regalis*, which grows along the edges of rivers, could be used as

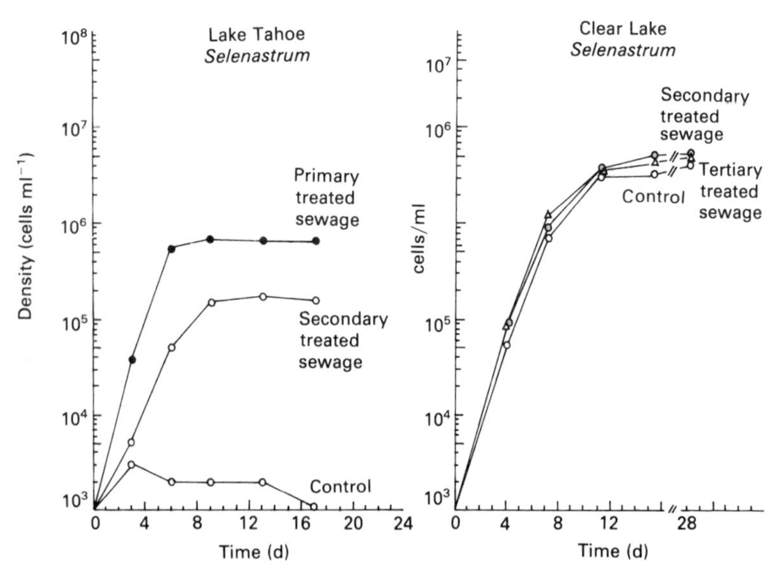

Fig. 7.3. The response of *Selenastrum capricornutum* to spikes of sewage in oligotrophic Lake Tahoe and eutrophic Clear Lake waters (from Payne, 1975).

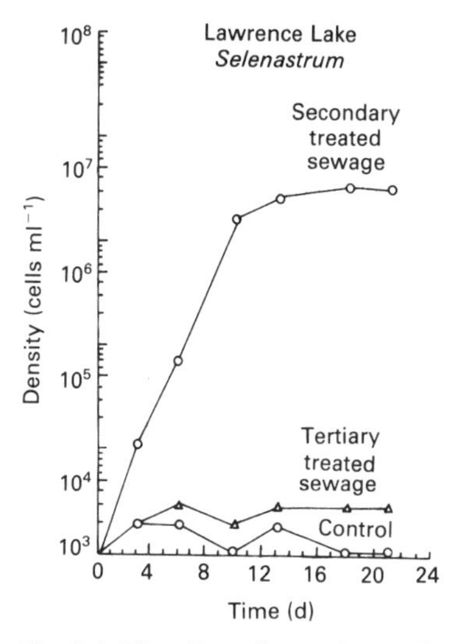

Fig. 7.4. The effect of secondary and tertiary sewage spikes on the growth of *Selenastrum capricornutum* in water from Lawrence Lake, Michigan (from Payne, 1975).

bio-assay material for detecting mutagens. Populations of ferns growing in a river heavily polluted with paper processing wastes had a high incidence of post-zygotic mutational damage, whereas neighbouring populations from non-polluted environments showed no mutational damage. *Osmunda regalis* is a long-lived perennial so that it may integrate the effects of mutational pollutants and detect chronic low doses or periodic high doses of mutagens.

Toxicity tests

The terminology used in toxicity testing and the responses of organisms to acute and chronic levels of toxic pollution have been described in Chapter 4. Brown (1976) lists twelve basic types of investigation into toxicity, most of them tests in the laboratory. The chief uses of toxicity tests are for a preliminary screening of chemicals, for monitoring effluents to determine the extent of risk of aquatic organisms and, for those effluents which are toxic, to determine which component is causing death so that it can receive especial treatment.

Studies of toxicity can be conducted in the field, using caged organisms, usually fish. Price (1979) described how brown trout (*Salmo trutta*) were caged in three Welsh lakes, one of which was receiving heavy metal pollution from mine wastes and should have been toxic to trout. Sudden changes in weather conditions were found to cause substantial mortality in the caged trout in the polluted lake, as large quantities of zinc were washed into the water, but at other times the water was not toxic to trout. Price pointed out that the use of caged fish in the field makes heavy demands on labour to examine and feed the fish, while vandalism is a recurring problem.

The simplest type of laboratory toxicity test is the static test, in which the organism is placed in a standard tank in the water under examination, for 48–96 h. There are normally a series of tanks with test water of different dilutions, usually in a logarithmic series. The organisms are removed at the end of the test period and total mortality is recorded. In these tests, the poison may evaporate, degrade or adsorb onto the surface of the tank, so that toxicity may be underestimated. These tests are nevertheless useful when an effluent needs rapid evaluation.

A more sophisticated method involves the periodic replacement of test water and an example is illustrated in Fig. 7.5, where 100 ml of solution is passed every 10 min. Four tubes are connected to the bottom of a 105 ml burette and each tube opens once every 10 min, while the others remain closed. A tube connecting the burette with a supply of dilution water opens to allow in a required amount of dilution. The tube connected to the air supply then opens to aerate the dilution water, after which the tube connecting the burette to the constant head vessel containing the test solution opens to allow the required volume into the burette, to bring the level up to 100 ml, which is then allowed to drain into the test vessel. This approach maintains the dissolved oxygen in the test vessel and removes excess carbon dioxide and ammonia. A standard dilution water is used and a test normally consists of ten dilutions, with controls of dilution water alone. The temperature is maintained at 20 °C.

Continuous flow systems are also used, but these require large volumes of clean water for dilution. A continuous flow system is illustrated in Fig. 7.6, where the diluent water reservoir (A) can provide water for 5 days. The diluent water flows, by gravity, from the reservoir to a constant head box (C) and then, at a constant rate, to a diluter, where mixing occurs with test water from a supply reservoir (D). The constant head box adjusts the temperature of the water, controlled by a thermostat in the test tank (E). The acclimating tank (B) allows the organisms to acclimatize to the diluent water before being placed in the test tank (E).

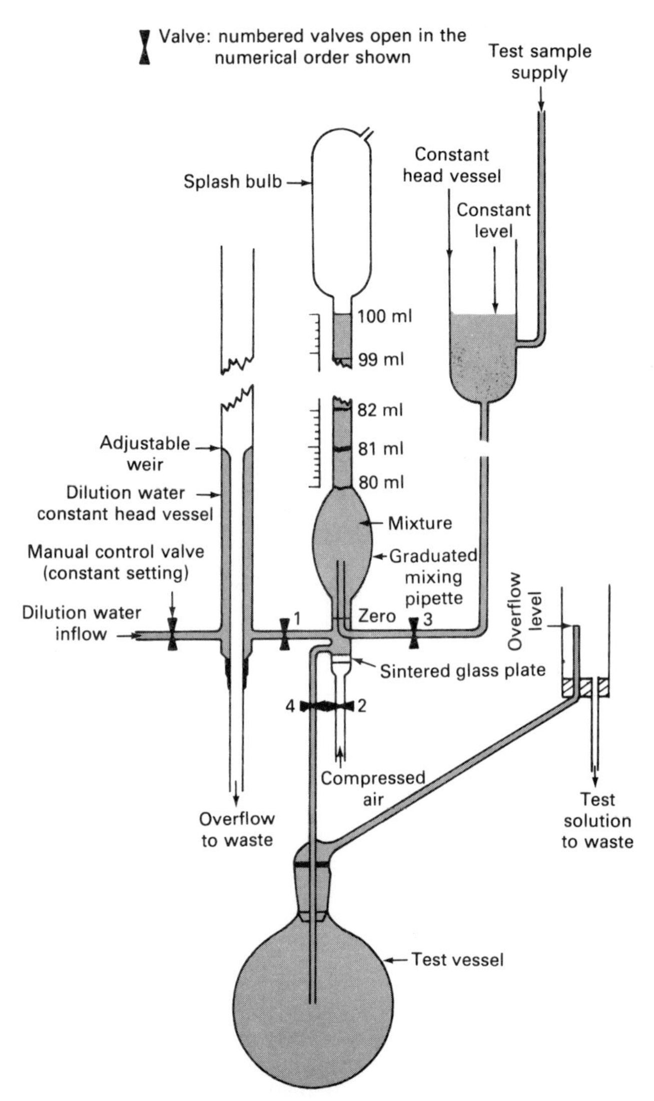

Fig. 7.5. Periodic replacement apparatus for standard toxicity tests using fish (from Muirhead-Thomson, 1971, after Alabaster and Abram, 1965).

Fish are normally used in standard toxicity tests. In the United States, fathead minnows (*Pimephales notatus*) and bluegill (*Lepomis macrochirus*), together with goldfish (*Carassius auratus*) and guppy (*Poecilia reticulata*) are the main test species, though there is a tendency now towards the use of a range of fishes. Early work in

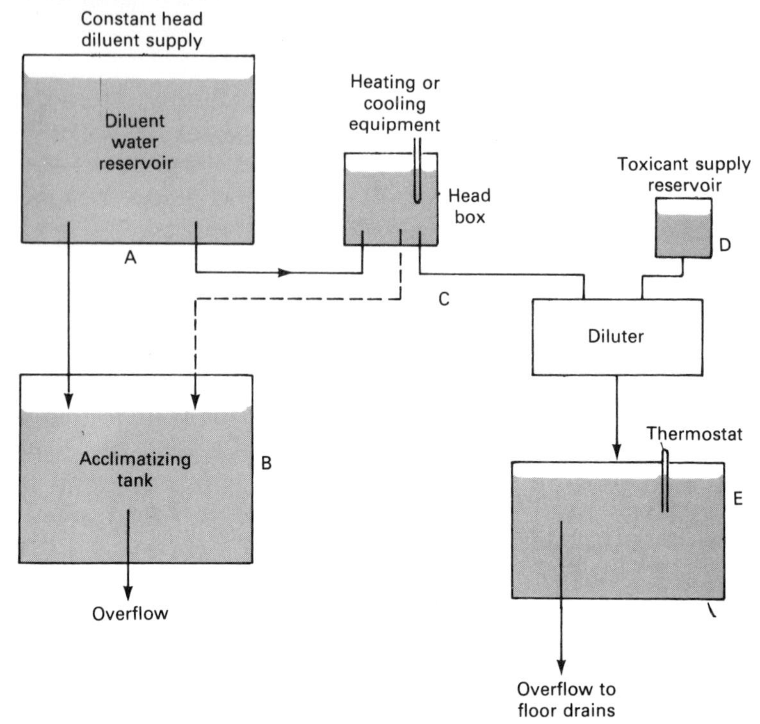

Fig. 7.6. A continuous flow system for toxicity tests (from American Public Health Association, 1976).

Britain was concentrated on brown and rainbow trout, but the tropical harlequin fish (*Rasbora heteromorpha*) has now largely superseded them in the laboratory because it is small and has a similar sensitivity to trout.

Much recent work has shown that the planktonic crustacean *Daphnia magna* could also be useful in routine work, being sensitive to pollutants, easily cultured and having a high reproductive rate (e.g. Adema, 1978; Leeuwangh, 1978; Tevlin, 1978). It has also been suggested recently that the luminescent bacterium *Photobacterium ficheri* might replace fish for some toxicity tests (LRE, 1979). The bacteria are inoculated into the water under test and the decrease in light production is proportional to the concentration of toxicant. The test takes minutes, rather than days, and could obviously prove invaluable where rapid decisions concerning effluents have to be made. Another new approach is the use of cultured mammalian cells for toxicity testing (Maruoka, 1978), which could be especially useful when screening for carcinogens.

Good general accounts of toxicity tests are given in Sprague (1969, 1970, 1971), Mirhead–Thomson (1971) and American Public Health Association (1976). For management purposes, the results of acute toxicity tests need to be related to field conditions in terms of safe concentrations of effluents and this has led to the introduction of application factors, many of which are purely arbitrary. In the United States, application factors are frequently very stringent. For example, for pesticides the application factor is 1/100 the 96 h lethal concentration threshold for the most sensitive life history stage (Hunter, 1978). A wide range of tests, ranging from biochemical to behavioural, have been developed to examine the effects of sublethal pollution (p. 109–115), but this very variety means that there has been little attempt at standardization, so that they are not yet used extensively in routine water quality monitoring for determining standards.

Fish alarm systems

Fishes show distinct physiological and behavioural responses to low levels of pollutants and attempts have been made recently to harness these in devising automatic alarm systems. An automatic fish monitor should provide a rapid indication that water quality has deteriorated and they have potential use in monitoring river waters and raw waters which are abstracted for potable supply and for monitoring effluents from sewage treatment works and industrial plants. Such monitors must be on line to the operations control centre so that immediate action can be taken if an alarm is sounded.

To be fully successful, a number of requirements must be met (Brown, 1978).

1. the detector system must always be alerted at such times as it is likely to be necessary;
2. the detector must be capable of responding unfailingly to the selected signal;
3. the detector must be sensitive enough to detect the signal at relevant levels;
4. the detector must always give the alarm when the selected level is exceeded;
5. the alarm signal should be a clear, characteristic and unequivocal one;
6. the alarm must be given soon enough so that any action required can be taken in time to be effective;
7. the system should not give false signals.

To date, fish alarm systems have monitored movement and respiratory activity in relation to levels of pollution. A typical tank, in which increasing opercular rhythms and movement of fish in relation to sublethal levels of pollutants are monitored, is illustrated in Fig. 7.7 (Westlake and Van der Schalie, 1977). To remove environmental effects, the experimental tanks are usually enclosed in a chamber, which reduces sound and vibrations and allows the photoperiod to be controlled, while water temperature is maintained constant. Food is given from an automatic feeder. Water is provided on a continuous flow basis. One fish is held per tank. The opercular movements of the fish result in a change of potential between electrodes. The movement of a fish can be recorded as it swims across a light beam being detected by two photocells and Morgan (1978) has used infra-red emitters instead of visible light so that the nocturnal rhythms of fish were not altered. Morgan (1979) has given a detailed description, with circuit diagrams, of a fish alarm system, in which photoelectric cells divide a unit into eight compartments to detail the movements of the fish. This system incorporated six tanks and water quality was considered to have deteriorated when four of the fish gave alarm responses.

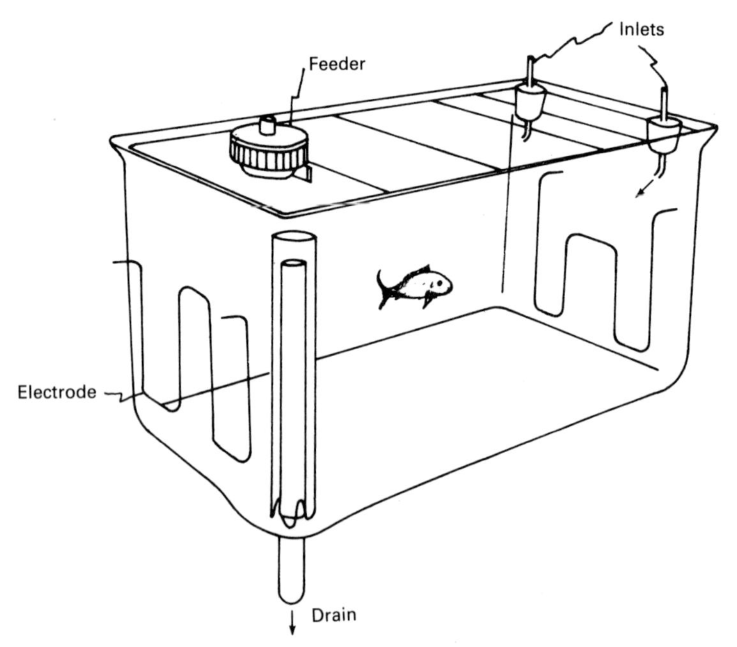

Fig. 7.7. An automatic fish monitor tank (from Westlake and Van der Schalie, 1977).

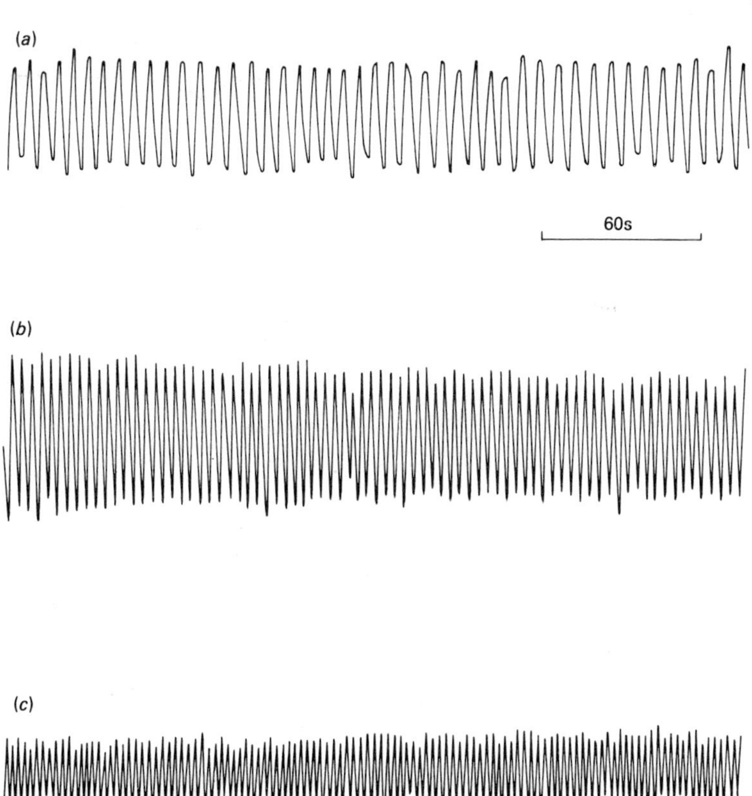

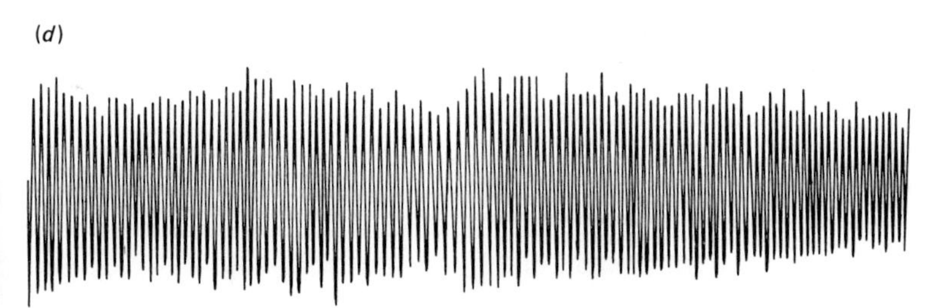

Fig. 7.8. The opercular rhythms of largemouth bass under natural conditions (a) and after exposure to 5.0 mg l^{-1} copper for 2 h (b), 4 h (c) and 60 min before death (d) Chart speed 100 mm min^{-1} (from Morgan and Kühn, 1974).

Besch *et al* (1974) and Scharf (1979) have used the capacity of fish to swim upstream and remain stationary in flowing water as the basis of an alarm system. The fish either flee or drift downstream in the presence of poisons. An electric shock keeps healthy fish away from the detector, which may be a beam of light.

An example of the type of results obtained is shown in Fig. 7.8, where the opercular rate of largemouth bass (*Micropterus salmoides*) are recorded after various periods of exposure to an eventually lethal dose of 5.0 mg l^{-1} copper, and it can be seen that the frequency of opercular movements increases. Carlson and Drummond (1978) have used the coughing frequencies of fish, (i.e. the reversal of water flow over the gills, which reduces the efficiency of oxygen uptake), which is similarly measured using electrodes, rather than opercular rate. The changes in cough rates of fish exposed to effluent from an ammunition plant are shown in Table 7.1.

Table 7.1. Percentage change in cough rate of bluegill sunfish (*Lepomis macrochirus*) after 24 h in an ammunition plant wastewater and control waters (after Carlson and Drummond, 1978)

Effluent concentration (%)	*Change in cough rate (%)*
1.0	1131*
0.75	1225*
0.56	774
0.32	258
0.18	182
0.06	143
Receiving stream water	91
Lake Superior water	49

* Fish died during test

Brown (1976) considered that a number of the criteria, listed above, for a fish alarm system were not met, so that alarms will not inevitably be given when harmful concentrations of pollutants are present and false alarms are liable to be given. Even when they give no alarm, this is no indication of the safety or suitability of the water for man or animals. Brown states that an alarm system which is not foolproof is no alarm system at all, thus throwing cold water on this developing field. Price (1979) shows that, for only two pollutants (copper and cyanide) out of ten, would a fish indicate within 24 h that a maximum acceptable concentration for that pollutant in potable water had been exceeded. With a response period of 24 h, a monitoring site would have to be considerably upstream of an abstraction point, so that any pollution discharged between the two localities would be missed. The value of such a monitor, however, is that it could detect a wide range of pollutants, at levels of a fraction of the concentration which proves lethal, and test water is under continuous surveillance. The development of this field is very recent and it may be that organisms other than fish may eventually prove more sensitive.

Chapter 8

THE WATER INDUSTRY

Management of the hydrological cycle

The prime task of the water industry is to manage the hydrological cycle for the benefit of users. The main sources of water are from reservoirs, or natural lakes, from river flows and from groundwaters, though other sources, e.g., sea water exploitation by desalination or rainmaking by cloud-seeding, may be important in some parts of the world. The water industry develops these resources and manages them to provide water, wherever possible, in the quantities required by users. Water will be required by domestic, industrial and agricultural users and must be of an acceptable quality for these purposes, requiring varying degrees of treatment.

Water is used extensively for recreation and amenity and the water industry is responsible for the development and regulation of these demands. Within this area are fisheries, many of which are of economic value, and in underdeveloped countries, fish may be the chief source of protein to the local community. Conservation of wild life resources and the landscape also fall within the general area of amenity.

As we have seen, society produces a vast array of waste products, water providing an effective means of disposing of many of these and the water industry is responsible for ensuring that the disposal of wastes causes the minimum of damage to resources. Water can also be the source of electric power, using dams and turbines. The water industry is also responsible for reducing the damage caused by flooding, which can be highly destructive of life, industry and agriculture. Navigation must also be maintained on many watercourses. The management system is summarized in Fig. 8.1.

Porter (1978) points out that water sources are often linked by management. The flow of rivers, for example, may be regulated by the controlled addition of water from reservoir or groundwater

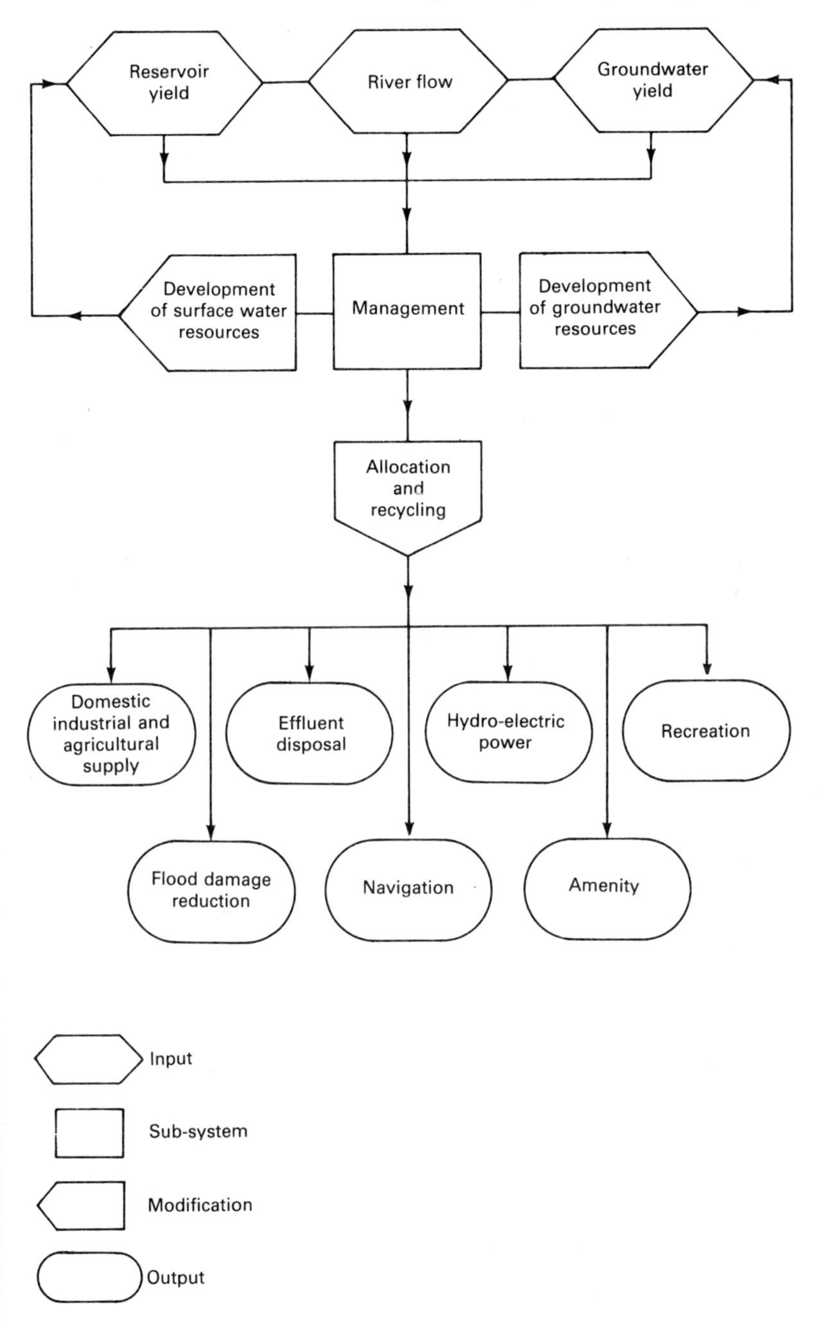

Fig. 8.1. The water management system (from Porter, 1978).

sources and water may be withdrawn from rivers at peak flows for storage in reservoirs or for refilling aquifers. Many of the outputs of management may be complementary but others may be in conflict. Land drainage and flood prevention schemes may, for instance, involve the destruction of prime wildlife habitats, a topic to be discussed later.

The water industry has little control over some key points in the water cycle, for instance precipitation and heavy floods. This is not to say that man's influence is not profound on these parts of the hydrological cycle. The destruction of forests, particularly those of the tropics, is a case in point (Myers, 1979). The forests generate much of the rainfall, up to 50 per cent in the case of the Amazonian rainforest, and much of this may be lost with the destruction of forests, a process occurring at a rate of 110 000 km^2 per annum in the tropics. Forests also act as sponges, absorbing water and regulating its release to rivers. The felling of forests results in rapid run-off, followed by drought, so that agricultural areas lower in the catchment alternately suffer devastating floods, followed by periods of water shortage, resulting in tremendous loss of life and agricultural production. Floods erode away valuable soil, depositing it in rivers and estuaries to damage fisheries and interrupt navigation. The large scale destruction of rainforests in the third world, frequently by multi-national organizations from the developed world seeking short-term profits, may also have major repercussions for world climate. More solar heat may be reflected from land cleared of forests, altering the global patterns of air circulation and wind currents, possibly decreasing rainfall in equatorial and temperate lands. Forests also act as a sink for carbon dioxide, but, with large-scale burning, they could become a source. An increase in atmospheric carbon dioxide could, through the greenhouse effect, increase the temperature of the earth and a rise of only 1 °C would, according to Bryson and Murray (1977) decrease the output of the productive American grain belt by 11 per cent. Obviously such global effects, were they to occur, would be catastrophic for the carefully managed water resources of the developed world, thus emphasizing that resource conservation must, in the long term, be viewed on a global scale. Some 50 per cent of the world's population is directly affected by the way watersheds are managed (Allen, 1980). Management of water resources must involve a consideration of the entire catchment, a topic described by Pereira (1973).

Figure 8.2 (shaded area) illustrates the main points of intervention in the water cycle. Intervention involves regulating the passage of

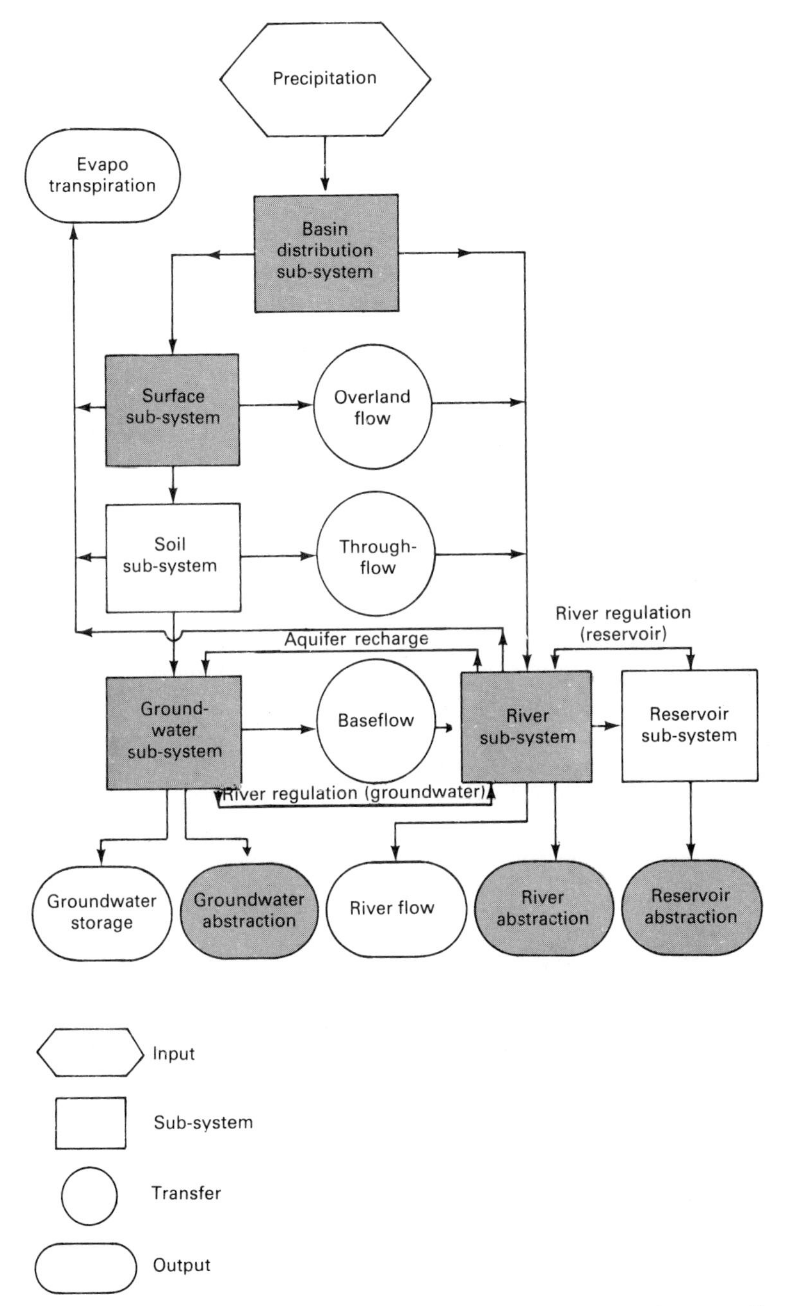

Fig. 8.2. Points of intervention by the water industry in the river basin system (from Porter, 1978).

water through different storages, diverting between storages and altering the retention time so that water is available when required. Availability is thus, to some extent, emancipated from the pattern of precipitation. Water is abstracted into supply from groundwater, rivers and reservoirs and these sources are used in conjunction with one another, water frequently being moved between sources and between regions to where it is required.

The industries of water supply, waste disposal, fisheries, etc may be managed separately, but for efficient resource management, all facets of the industry should ideally be under single control. This is effectively what now happens in England and Wales with the reorganization of the water industry following the 1973 Water Act. Ten regional water authorities, which resemble nationalized industries, were set up and given control over water supply, effluent disposal and recreation and amenity. A number of water companies, supplying water to domestic and industrial users, are still extant, acting as agencies for the water authorities. The water authorities are based on a catchment, or series of catchments, so that overall watershed management is practicable (Fig. 8.3). A national body, the National Water Council, of which the chairmen of all the water authorities are members, provides a national overview of water resource management, as well as taking overall responsibility for training and salaries. A national policy for developing water-based recreation is provided by the Water Space Amenity Commission. The Water Research Centre was set up to promote, assist and support research relating to the management of the water cycle. Plans are afoot to combine the National Water Council, the Water Research Centre and some of the data-gathering functions of the Department of the Environment into a single National Water Authority. Government departments, chiefly the Department of the Environment, the Welsh Office and the Ministry of Agriculture, Fisheries and Food ensure that a national policy for water is effectively executed. The organization and functioning of the water industry in Britain is described in Okun (1977), Sewell and Barr (1978) and Dangerfield (1979).

Biology and the water industry

The construction of reservoirs, the drainage of land, the building of sewage works etc., are largely the provinces of engineers and the water industry developed with a domination of engineers, initially to the exclusion of other disciplines. Many large-scale works have been

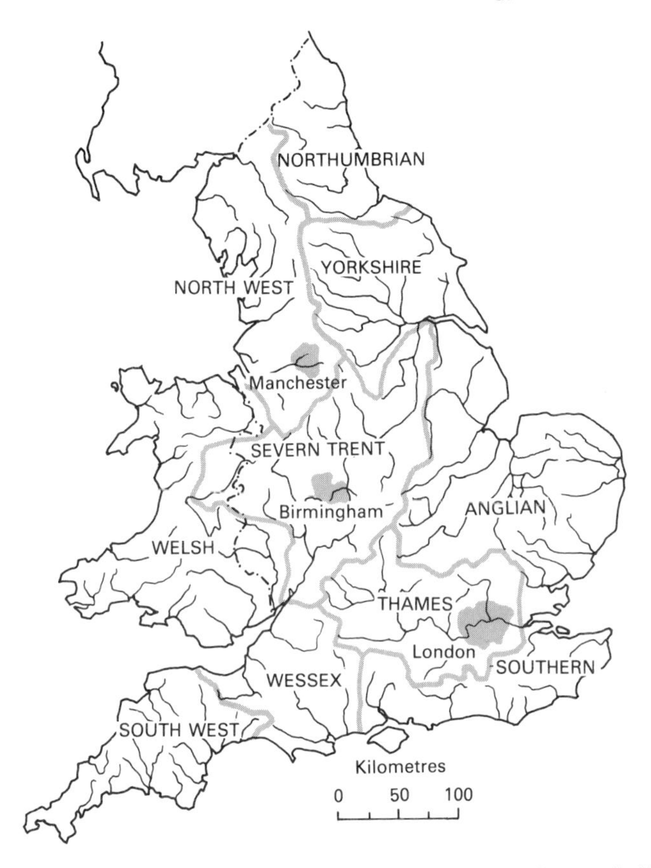

Fig. 8.3. The Water Authorities, in relation to river catchments, in England and Wales.

executed in the past without advice on the ecological consequences, which have frequently been severe. Large reservoirs today are usually designed with multiple use in mind, i.e., as well as the generation of electricity or the storage of water, reservoirs can also be used for irrigation, fisheries, controlling floods and recreation. Multiple use can help mitigate some of the adverse factors consequent on the building of a reservoir.

Goodland (1977) has described the construction of the Bayano hydroproject, for the generation of electricity, in the Panama isthmus of central America, in which scant attention was paid to the ecological consequences. The 70 m high dam will impound a lake 300 km long, with an area of 65 000 ha. The land has been flooded without removing the tropical rain-forest, which could have been harvested at a profit of 700 000 US dollars. The slow decomposition of these trees will provide nutrients, causing an explosive growth of aquatic macrophytes, which provide an ideal habitat for disease vectors, such as mosquitoes and snails. The tops of the trees will be exposed during maximum drawdown and these will provide hazards to navigation and make fishing and weed control difficult, if not impossible, in some places. The Bayano reservoir has flooded valuable, fertile agricultural land. In a multiple-use plan, this could have been at least partially offset by developing a fishery. The leaves etc., of the drowned forest will decompose to cause eutrophication, anaerobic situations and possibly toxic conditions, whilst the trunks will stand indefinitely, impeding the use of nets. As many fish species feed in the shallows, the proposed drawdown of the lake will seriously affect any development of a fishery. The Bayano river, the largest of the five flowing into the Gulf of Panama, carries a silt load which fertilizes a shallow bay, supporting a valuable industry for fish and shrimps, the latter valued at 2 per cent of Panama's GNP. The dam will prevent fertilization of the estuary by Bayano silt and could reduce the fishery by as much as 15 per cent. The Aswan Dam in Egypt and the Kariba Dam in Tanzania provide other well-known examples of projects which, without adequate ecological planning, have produced inordinately expensive problems.

The biologist in the water industry

The domination of the water industry by engineers meant that few biologists were originally employed, but the realisation of the importance of biology to the management of water resources, together with the re-organization of the industry in England and Wales, has resulted in a rapid increase in the employment of biologists. For example, in England and Wales, the industry employed fewer than 30 biologists prior to 1960 whereas in 1976 there were some 350 (Leeming, 1978).

There are a number of objectives in the management of water quality (UNESCO/WHO, 1978) in which the biologist plays an important part.

(a) *The classification of water resources.* Water resources are increasingly classified according to the uses to which they are put and the quality necessary for a particular use must be defined. For example, certain industrial processes, raw water for potable supply and game fisheries require water of the highest quality, whereas low quality water is acceptable in a waterway used primarily as a receiving body for waste effluents. An assessment of the biological resources is necessary in an initial classification.

(b) *The collection of baseline data*. These data will allow any changes in water quality caused by the development of a resource to be detected, together with changes which may interfere with the present and planned usage of water.

(c) *Water quality surveillance*. This is routine work carried out to determine the effectiveness of waste-water management programmes or carried out in relation to specific uses, such as water abstraction, fisheries or recreation.

(d) *Specific investigations*. These may involve determining the effects of a specific pollution incident and the subsequent recovery of the freshwater community, the effect of a new impoundment on water resources downstream, the development of communities within impoundments etc., i.e., any investigation in relation to a specific or potential problem.

(e) *Forecasting*. The forecasting of changes due to a variation in the intensity of use of a resource or to altering pollution inputs is essential for the rational exploitation of water resources. The biologist can provide essential input for the development of predictive models for forecasting both in the short term and the long term.

For the tasks outlined above the biologist will ideally be working as a member of a multifunctional team, but there are two further areas in which the biologist plays a dominant role:

(f) *Fisheries*. The development and maintenance of commercial and recreational fisheries is an essential function of the water industry. Interest in recreational fishing has expanded very rapidly over the last two decades and, in Britain, angling has become the greatest single participating sport. The development and optimization of fisheries resources and their maintenance in relation to other conflicting uses of water requires a high level of expertise. In many cases fisheries management is kept separate from other biological work but ideally a multifunctional team of biologists, including those primarily involved in pollution assessment, is the most rational way of managing fisheries. The types of studies carried out by fisheries biologist within the water industry have been listed by Leeming (1978) as:

   (i) The biological evaluation of the habitat, in conjunction with water quality studies, and the assessment of suitability for particular species of fish.

(ii) The growth and production of fish stocks in various habitats.
(iii) Investigation of the effects of land drainage works on river fisheries.
(iv) The production of fish in culture for the stocking of rivers and lakes.
(v) The movement and migration of migratory species.
(vi) Fish diseases and pathology.

(g) *Wild-life conservation*. The biologist within the water industry is normally not involved in the wider aspects of wildlife conservation, being primarily concerned with occurrences within rivers and lakes, but the management of water resources frequently produces changes which profoundly affect sensitive habitats within a catchment, an area to be dealt with in the next section.

Water quality standards

Management decisions are based on the comparison of water quality data with criteria and standards. Such standards, relating to uses of water, are reasonably well developed in the physical and chemical fields, but biological standards are still in an early stage of development. The following definitions (from UNESCO/WHO 1978) are generally accepted:
Criteria: a scientific requirement on which a decision or judgement may be based concerning the suitability of water quality to support a designated use.
Objectives: a set of levels of water quality parameters to be attained in water quality management programmes, which also involve cost/benefit considerations.
Standards: legally prescribed limits of pollution which are established under statutory authority.

Standards and their derivation are discussed in general by Holdgate (1979) and for freshwaters by Price and Pearson (1979). The practical management of a river has been summarized by Toms (1975) as consisting of six stages:

(a) To decide on the uses of a particular river.
(b) To decide on the water quality conditions necessary in a river in order to support the uses decided under (a).
(c) To assess the effect of existing discharges on a river and to attempt a forecast of future effluents.
(d) To decide upon the standards that are required for each effluent discharge in order to leave a river of the necessary quality.

(e) To produce consent conditions which will include the standard decided under (d).
(f) To initiate a sampling programme which will both ensure that discharges comply with the above standards and also indicate from the river water quality whether revised effluent standards are necessary.

Wildlife conservation and the water industry

The biological resources of immediate concern to the water industry are fishes and their food, together with those organisms of use in detecting changes in water quality. The work of a water authority, however, affects a far greater range of habitats and their biota than merely the watercourse itself. Where a large-scale engineering scheme is planned an environmental impact assessment will take account of the biological changes which may result and the overall desirability of the scheme can be fully debated. In terms of wildlife conservation it is the small scale destruction of habitat, over a wide area, which is proving so detrimental. While not definable as pollution, habitat modification by the water authority can be as damaging to the biota as pollution and hence should be the concern of the biologist within the industry. In Britain, Section 22 (i) of the Water Act 1973 states:

In formulating or considering any proposals relating to the discharge of any of the functions of the Water Authorities, those authorities and the appropriate Minister shall have regard to the desirability of preserving natural beauty, of conserving flora, fauna and geological or physiological features of special interest, and of protecting buildings and other objects of architectural, archeological or historic interest and shall take into account any effect which the proposals would have on the beauty of, or amenity in, any rural or urban area or on any such flora, fauna, feature, buildings or objects.

The drainage of land for agriculture and channel improvement to aid flood prevention cause most concern to the wildlife biologist. The discussion below will be confined to the situation in Britain.

Some 100 000 ha of land were drained in England and Wales during the year 1978/1979, compared with 51 000 ha in 1968/69, 30 000 ha in 1958/59 and 20 000 ha in 1951/52. According to Newbold (1977), there are about 162 500 km of watercourses in Britain, including 3200 km of canals, 30 600 km of main river, 32 000 km of smaller watercourses and 96 700 km of ditches and dikes. The Regional Water Authorities manage some 30 per cent of rivers, while the smaller watercourses are managed by 267 Internal Drainage Boards, whose membership is made up mainly of farmers and landowners, their voting strength depending on the percentage of land they own. IDB members thus have a vested interest in land drainage for agricultural improvement. Land drainage is heavily subsidised by government, to the tune of £13.2 m for arterial drainage in 1976/77. It is of interest that the *total* government grant in 1978/79 to wildlife conservation, via the Nature Conservancy Council, was a derisory £7 m, compared with £540 m to the Ministry of Agriculture, Fisheries and Food. It is small wonder that conservationists are powerless to prevent the loss of many prime wetland sites and are probably unaware that many small but important wetlands are disappearing.

The drainage of wetlands results in the loss of species rich grasslands, fens and marshes to species poor, improved grasslands or ultimately to arable monocultures. Of plant species in Britain which have become extinct, rare or are rapidly declining, 22 per cent have been adversely affected by land drainage (Newbold, 1977). The attractive fritillary, *Fritillaria meleagris*, once a widespread plant in lowland English meadows, has become restricted to a few nature reserves in the Thames Valley and Suffolk, while the snipe, *Gallinago gallinago* a wading bird breeding in wetlands, has become extinct over much of lowland England (Mason and Macdonald, 1976). A localized drainage scheme may affect the water table over a considerably greater area so that a landowner, interested in preserving a wetland, may find his habitat deteriorating due to the activities of his neighbours.

A typical example of wetland destruction is the decision by Government in 1979 to grant aid the drainage of some 16 ha of West Sedgemoor, a Site of Special Scientific Interest, adjacent to a nature reserve, despite objections by the Nature Conservancy Council and voluntary conservation bodies. Drainage of such sites is carried out ostensibly in the national interest, because Britain imports large quantities of food. Nevertheless Britain, as part of the European Economic Community, produces quantities of food in surplus, which is later sold at considerable loss to eastern bloc countries. Government grants at present are used to drain land often for growing crops already in surplus or adding to the dairy products' stockpiles of the EEC. Furthermore, large areas of eastern England, once prime wetland areas, are used for growing bulbs and cut flowers rather than foodstuffs. One can only conclude that the majority of land drainage involving ecologically valuable sites is carried out to maximize the profits of individual land owners rather than to benefit the nation. The majority of these sites, once destroyed, can never be recreated so it would seem prudent, in the national interest, to conserve their ecological richness at least until the need for increased food production is shown to be critical.

The other main area of conflict between wildlife conservationists and the water industry is river management for flood prevention. Management involves the destruction of aquatic macrophytes by manual, mechanical or herbicidal techniques and the removal of bankside vegetation, particularly mature trees. The majority of Britain's river banks were originally wooded, but now many are devoid of bankside trees. In eastern England up to 70 per cent of bankside trees were removed between 1879 and 1970, 20 per cent in the period 1960 to 1970 (C. F. Mason and S. M. Macdonald, unpublished data). Figure 8.4 shows the level of management which takes place and this will obviously have a serious effect on wildlife. The otter (*Lutra lutra*), an amphibious carnivore, has seriously declined over England and Wales, and indeed most of western Europe, over the last few decades and is now endangered (Macdonald and Mason, 1976; Nature Conservancy Council, 1977). The marked decline of the otter has been correlated strongly with the widespread introduction of persistent pesticides in the late 1960s (Chanin and Jefferies, 1978), but the failure of the species to recover, with the withdrawal from the market of many of these compounds, appears to be due to the attrition of habitat. Otters lie up and breed in bankside dens, and the root systems of mature trees, especially oak (*Quercus* spp.), ash (*Fraxinus excelsior*) and sycamore (*Acer pseudoplatanus*) are especially favoured. These are the very trees removed by water engineers and such den sites for otters are now at a premium (Macdonald *et al*, 1978). The otter situation is reviewed by Macdonald and Mason (1980).

Recent work has suggested that current management practices are excessive even for effective flood prevention. Krause (1977) has shown in West Germany and the Netherlands that bankside tree cover reduces the growth of macrophytes within the river and hence the need for their removal. Tree removal is also expensive and, by leaving them, overall maintenance costs are reduced by 50 per cent. Dawson (1978) has suggested that, while the removal of macrophytes temporarily reduces the risk of flooding, subsequent plant growth is both vigorous and synchronous and that the maximum biomass reached in managed waters is frequently greater than in unmanaged

Fig. 8.4. Management of rivers in England.
(a) Unmanaged, with dense growth of mature trees overhanging the river, providing good wildlife habitat.
(b) Lightly managed. Trees are vigorously re-growing and there is a rich fen vegetation fringing the river, providing good wildlife habitat.
(c) Heavily managed and canalized, with trees and fringing vegetation eliminated, of no value to wildlife. (Photographs, Vincent Weir).

waters. Dawson recommends the planting of trees to reduce light as a good ecological control of aquatic plants and if the tree lines were incomplete, to allow intermittent growth of macrophytes, then an ideal solution to the needs of water engineers, fisheries managers and wildlife conservationists could be achieved.

Considerable conflict exists between water engineers and wildlife conservationists, not least because the latter do not have the resources to monitor the effects, potential and actual, of engineering schemes. The water industry has a statutory duty to have due regard for conservation and ideally the very minimum recognition of this duty would be the employment by each water authority of a wildlife biologist to examine the potential effects of any planned engineering or maintenance scheme and to liaise with statutory and voluntary conservation bodies. At comparatively small cost relative to total resources the conservation of wildlife as part of overall watershed management could be placed on a much sounder base.

Conclusions

The management of water resources and the control of pollution is expensive and the benefits of a particular remedial action must be carefully weighed against the cost. The public, in general, are showing growing concern over pollution. Studies in New York State, for example, showed that 98 per cent of people interviewed considered that the quality of the environment could be improved. Water pollution was thought to be a special problem and the majority insisted that pollution control was ineffective and that legal sanctions against polluters should be taken. One third of respondents thought that the cost of pollution control was high, but three quarters stated that all available pollution control techniques should be applied, even considering the cost, thus emphasizing the public's concern over pollution (Gore *et al*, 1975; Porter, 1975). The water industry clearly has a mandate to control pollution, but the public has a right to assure itself that its money is being wisely spent, which requires access to information generated within the industry. However, the water industry appears particularly indifferent to, or inept at, accounting for its actions to the public.

The control of water pollution is proving effective in many parts of the world. In England and Wales, for example, with a high population and large areas of intensive agriculture and urbanization, the Department of the Environment (1978) could claim that some 76 per cent of watercourses were free of organic pollution in 1975 and that there had been a substantial improvement in conditions over the previous decade. Holdgate (1979) takes an optimistic view of the threat of pollution and points out that deaths caused by pollution have been rare, very considerably less than mortality caused by accidents on the roads or in the home, or than self-inflicted damage caused by smoking and drinking. Nevertheless, newly synthesised materials are constantly being added to our waterways as traces in effluents, and the long-term effects of pollutants, acting alone or in combination, are still largely unknown. That populations of animals have been decimated over wide areas, by pollutants previously unknown to be environmental contaminants, should warn against complacency. Constant vigilance is required to protect the environment and to effectively manage water resources and these functions will require the expertise of biologists.

METHODS OF DERIVING BIOTIC AND DIVERSITY INDICES

The data from a lowland river, given on p. 210, are used in the calculations of Appendices A–D.

A. Trent biotic index (TBI) Refer to the Table on pages 202 and 203

Determine first the number of Groups present in the sample from the lower section of the Table (p. 203).

For the sample: Platyhelminthes (1) + Annelida (1) + Nais (1) + Hirudinae (3) + Mollusca (6) + Crustacea (2) + Plecoptera (0) + Ephemeroptera (2) + *Baetis rhodani* (1) + Trichoptera (3) + Megaloptera (1) + Chironomidae (1) + *Chironomus* (1) + Simulidae (1) + other fly larvae (0) + Coleoptera (1) + Hydracarina (1) = 26 groups.

Now turn to the top of the Table (p. 202). As there are 16+ groups present, use the final column.

Plecoptera were not recorded, so progress to row 3.

There were two species of Ephemeroptera present (excluding *Baetis rhodani*, which is more tolerant of pollution than other may flies). The Biotic Index is found by turning to column 5 in row 3.

TBI = 9

Plecoptera do not normally occur in lowland rivers in eastern England so that this site can be considered free of pollution.

The Trent Biotic Index

Clean

Organisms in order of tendency to disappear as degree of pollution increases		*Column.*	*Total number of groups present*				
			1.	*2.*	*3.*	*4.*	*5.*
			0–1	*2–5*	*6–10*	*11–15*	*16+*
Row			*Biotic index*				
1	Plecoptera larvae present	More than one species	–	7	8	9	10
2		One species only	–	6	7	8	9
3	Ephemeroptera larvae present	More than one species*	–	6	7	8	9
4		One species only*	–	5	6	7	8
5	Trichoptera larvae present	More than one species†	–	5	6	7	8
6		One species only†	4	4	5	6	7
7	*Gammarus* present	All above species	3	4	5	6	7
8		absent					6
9	*Asellus*	All above species absent	2	3	4	5	
10	Tubificid worm and/or Red Chironomid larvae present	All above species absent	1	2	3	4	–
11	All above types absent	Some organisms such as *Eristalis tenax* not requiring dissolved oxygen may be present.	0	1	2	–	–

Polluted

* *Baetis rhodani* excluded

† *Baetis rhodani* (Ephem.) is counted in this section for the purpose of classification

The term 'Group' used for purpose of the biotic index means any one of the species included in the following list of organisms or sets of organisms:

Each known species of Platyhelminthes (flatworms)
Annelida (worms excluding genus *Nais*)
Genus *Nais* (worms)
Each known species of Hirudinae (leeches)
Each known species of Mollusca (snails)
Each known species of Crustacea (hog louse, shrimps)
Each known species of Plecoptera (stone-fly)
Each known genus of Ephemeroptera (may-fly, excluding *Baetis rhodani*)
Baetis rhodani (may-fly)
Each family of Trichoptera (caddis-fly)
Each species of Megaloptera larvae (alder-fly)
Family Chironomidae (midge larvae except *Chironomus riparius*)
Chironomus riparius (blood worms)
Family Simulidae (black-fly larvae)
Each known species of other fly larvae
Each known species of Coleoptera (Beetles and beetle larvae)
Each known species of Hydracarina (water mites)

B. Chandler biotic score

The Chandler Biotic Score

Row	Groups present in sample		*Abundance class in standard sample*				
			1	*2*	*3*	*4*	*5*
			Present 1–2	*Few 3–10*	*Common 11–50*	*Abundant 51–100*	*Very abundant 100+*
			Points scored				
		Crenobia alpina					
1.	Each species of	Taenopterygidae, Perlidae, Perlodidae, Isoperlidae, Chloroperlidae	90	94	98	99	100
2.	Each species of	Leuctridae, Capniidae, Nemouridae (excluding *Amphinemura*)	84	89	94	97	98
3.	Each species of	Ephemeroptera (excluding *Baetis*)	79	84	90	94	97
4.	Each species of	Cased caddis, Megaloptera	75	80	86	91	94
5.	Each species of	*Ancylus*	70	75	82	87	91
6.	–	*Rhyacophila* (Trichoptera)	65	70	77	83	88
7.	Genera	*Dicranota*, *Limnophora*	60	65	72	78	84
8.	Genus	*Simulium*	56	61	67	73	75
9.	Genera of	Coleoptera, Nematoda	51	55	61	66	72
10.	–	*Amphinemura* (Plecoptera)	47	50	54	58	63
11.	–	*Baetis* (Ephemeroptera)	44	46	48	50	52
12.	–	*Gammarus*	40	40	40	40	40

The Chandler Biotic Score—continued

Row	Groups present in sample		*Abundance class in standard sample*				
			1	*2*	*3*	*4*	*5*
			Present 1–2	*Few 3–10*	*Common 11–50*	*Abundant 51–100*	*Very abundant 100+*
			Points scored				
13.	Each species of	Uncased caddis (excl. *Rhyacophila*)	38	36	35	33	31
14.	Each species of	Tricladida (excluding *c. alpina*)	35	33	31	29	25
15.	Genera of	Hydracarina	32	30	28	25	21
16.	Each species of	Mollusca (excluding *Ancylus*)	30	28	25	22	18
17.	–	Chironomids (excl. *C. riparius*)	28	25	21	18	15
18.	Each species of	*Glossiphonia*	26	23	20	16	13
19.	Each species of	*Asellus*	25	22	18	14	10
20.	Each species of	Leech (excl. *Glossiphonia*, *Haemopsis*	24	20	16	12	8
21.	–	*Haemopsis*	23	19	15	10	7
22.	–	*Tubifex* sp.	22	18	13	12	9
23.	–	*Chironomus riparius*	21	17	12	7	4
24.	–	*Nais*	20	16	10	6	2
25.	Each species of	air breathing species	19	15	9	5	1
26.		No animal life			0		

B. Chandler biotic score Refer to the Table on pages 202 and 203.

Crenobia alpina and stoneflies (Plecoptera) are absent, so the table is entered at row 3, Ephemeroptera.

Row	*Number of taxa*	*Abundance class*	*Score*
3	2	2 + 3	79 + 84
4	2	1 + 1	75 + 75
5	0	–	–
6	0	–	–
7	0	–	–
8	1	4	73
9	1	2	55
10	0	–	–
11	1	3	48
12	1	4	40
13	2	2 + 3	36 + 35
14	1	2	33
15	1	3	28
16	4	1 + 2 + 3 + 3	30 + 28 + 25 + 25
17	1	4	18
18	1	3	20
19	1	5	10
20	2	2 + 3	20 + 16
21	0	–	–
22	1	4	12
23	1	5	4
24	1	3	10
25	0	–	–

Chandler Biotic Score = 879
Averaged Chandler Score = 879/24 = 37

The score indicates water largely free of pollution, though the absence of stoneflies in lowland rivers has given a score on the low side for clean water. In an upland river with a diverse stonefly fauna, scores can frequently approach 2000.

	Relative abundance classes			
	1	*2*	*3*	*4*
	1 %	*1–5* %	*6–20* %	*21–100* %
Groups	*Scores*			
1. *Crenobia alpina*, Taenopterygidae, Perlidae, Chloroperlidae	60	90	95	100
2. Perlodidae	50	85	90	95
3. Leuctridae, Capniidae	50	85	90	95
4. Nemouridae, Ephemeroptera (except *Baetis* spp.)	40	70	75	80
5. Cased Trichoptera, Rhyacophilidae, *Dicranota* spp., *Limnophora* spp., *Atherix* spp., *Ancylus fluviatilis*	25	55	60	65
6. Uncased Trichoptera (except Rhyacophilidae)	20	45	55	60
7. *Baetis* spp	15	45	50	55
8. *Gammarus* spp., Hydracarina, Tricladida (except *Crenobia alpina*)	15	40	40	40
9. Mollusca (except *Ancylus fluviatilis*), Chironomidae (except red Chironomini), Hirudinea	10	30	25	20
10. *Asellus* spp.	10	20	15	10
11. Naididae, Red Chironomini	5	20	10	5
12. Tubificidae	5	10	10	5
13. Air breathing Diptera (e.g. *Eristalis* spp.)	0	5	5	0
14. Representatives of all the above 'groups' absent			0	

C. Group score index Refer to the Table on page 205

There were no *Crenobia* or stoneflies present, so enter table at Group 4.

Group	*Abundance class*	*Score*
4	1	40
5	1	25
6	2	45
7	2	45
8	3	40
9	4	20
10	3	15
11	3	10
12	3	10
13	0	–
		Group score = 250

This is only 39% of the maximum score of 645 and would indicate slight organic pollution at the site. As with the Chandler Score, the absence of stoneflies means that lowland rivers will obtain scores on the low side of the 'clean' range.

D. Calculation of species diversity

$$H' = -\sum_{n-1}^{s} pi \log pi$$

$pi = n_i/N = n_i/758$ for this sample

e.g.

	pi	$pi \log pi \times 10^{-3}$
Polycelis tenuis	0.00395	2.3566
Tubificidae	0.06860	125.9718
Nais elinguis	0.04089	65.8990
Glossiphonia complanata	0.01583	18.9878
etc. for all taxa.		

H' is the sum of $pi \log pi$ over all taxa
$= 2.55$

Wilhm and Dorris (1968) considered clean waters to have diversity values greater than 3, so the present station would appear to be slightly polluted.

Macro-invertebrates collected from the upper stretch of river in lowland eastern England in October 1979 (sum of 5 two-minute kick samples).

Phylum	*Class/Order*	*Family*	*Species*	*No. collected*
Platyhelminthes	Turbellaria	Planariidae	*Polycelis tenuis*	3
Annelida	Oligochaeta	Tubificidae	–	52
		Naididae	*Nais elinguis*	31
	Hirudinea	Glossiphoniidae	*Glossiphonia complanata*	12
			Helobdella stagnalis	9
		Erpobdellidae	*Erpobdella octoculata*	4
Mollusca	Gastropoda	Valvatidae	*Valvata piscinalis*	14
		Hydrobiidae	*Bithynia tentaculata*	1
		Lymnaeidae	*Lymnaea pereger*	11
		Planorbidae	*Planorbis vortex*	9
	Bivalvia	Sphaeriidae	*Sphaerium* sp.	22
			Pisidium sp.	45
Arthropoda	Crustacea	Asellidae	*Asellus aquaticus*	102
		Gammaridae	*Gammarus pulex*	61
	Hydracarina	Elayidae	*Eylais hamata*	37
Insecta	Ephemeroptera	Baetidae	*Baetis rhodani*	22
		Caenidae	*Caenis robusta*	2
		Ephemeridae	*Ephemera danica*	3
	Odonata	Coenagriidae	*Enallagma cyathigerum*	1
	Hemiptera	Corixidae	*Sigara falleni*	13
	Coleoptera	Elminthidae	*Elmis aenia*	7
	Trichoptera	Hydropsychidae	*Hydropsychae angustipennis*	33
		Polycentropidae	*Cyrnus trimaculatus*	5
		Limnephilidae		3
	Megaloptera	Sialidae	*Sialis lutaria*	3
	Diptera	Chironomidae		
			Chironominae 'red chironomids'	125
			Orthocladinae 'green chironomids'	56
		Simulidae	*Simulium* sp.	72
			Total	758

REFERENCES

Adema, D. M. M. (1978) *Daphnia magna* as a test animal in acute and chronic toxicity tests, *Hydrobiologia* **59,** 125–34.

Aho, J. M. Gibbons, J. W. and Esch, G. W. (1976) Relationship between thermal loading and parasitism in the mosquitofish. In Esch G. W. and McFarlane, R. W., *Thermal Ecology II* pp. 213–8, Technical Information Centre, Springfield, Virginia.

Alabaster, J. S. (1962) The effects of heated effluents on fish, *Int. J. Air Water Pollut.* **7,** 541–63.

Alabaster, J. S. and Abram F. S. H. (1965) Development and use of a direct method of evaluating toxicity to fish. *Advances in Water Pollution Research* Proc. 2nd Int. Conf., Tokyo 1964. Vol 1. pp. 41–54. Pergamon Press, Oxford

Alderdice, D. F. (1967) The detection and measurement of water pollution – biological assays, *Canada Dept. Fisheries: Can. Fish. Rept. no. 9*, 33–9.

Allen, R. (1980) *How to save the world*, Kogan Page, London.

American Public Health Association (1971) *Standard methods for the examination of water and waste water*, APHA, Washington.

American Public Health Association, American Waterworks Association, Water Pollution Control Federation (1976) *Standard methods for the examination of water and wastewater*, 14th edn, APHA, Washington.

Anderson, P. D. and S. D'Apollonia (1978) Aquatic animals. In Butler, G. C., *Principles of ecotoxicology* pp. 187–221. Wiley, Chichester.

Anderson, T. W. (1960) *An introduction to multivariate statistical analysis*, Wiley, Chichester.

Andersson, G., Cronberg, G. and Gelin, C. (1973) Planktonic changes following the restoration of Lake Trummen, Sweden, *Ambio* **2,** 44–7.

Anon. (1979) River rescue, *Wat. Waste Treatment* **22,** 38–42.

Archibald, R. E. M. (1972) Diversity of some South African diatom associations and its relation to water quality, *Water Res.* **6,** 1229–38.

Arthur, D. R. (1972) Katabolic and resource pollution in estuaries. In Cox, P. R. and Peel, J. (eds), *Population and pollution* pp. 65–83, Academic Press, London.

Aston, R. J. (1973) Tubificids and water quality: a review, *Environ. Pollut.* **5,** 1–10.

Austin, E. P. (1979) First-stage treatment. In *Water pollution control technology*, pp. 37–47, HMSO, London.

Baer, J. G. (1971) *Animal parasites*, Weidenfeld and Nicolson, London.

Balloch, D., Davies, C. E. and Jones, F. H. (1976) Biological assessment of water quality in three British rivers: the North Esk (Scotland), the Ivel (England) and the Taf (Wales), *Wat. Pollut. Control* **75,** 92–114.

Bardach, J. E., Fujiya, M. and Holl, A. (1965) Detergents: effects on the chemical senses of the fish *Ictalurus natalis* (Le Sueur), *Science, N. Y.* **148,** 1605–7

Bartha, R. and Atlas, R. M. (1977) The microbiology of aquatic oil spills, *Adv. appl. Microbiol.* **22,** 225–66.

Beak, T. W., Griffing, T. C. and Appleby, A. G. (1973) Use of artificial substrate samplers to assess water pollution, *Biological methods for the assessment of water quality*, *ASTM STP 528* pp. 227–41, American Society for Testing and Materials.

Beatson, C. G. (1978) Methaemoglobinaemia – nitrates in drinking water, *Environ. Health* **86,** 21–35.

Beattie, J. H. and Pascoe, D. (1978) Cadmium uptake by rainbow trout, *Salmo gairdneri*, Richardson, eggs and alevins, *J. Fish. Biol.* **13,** 631–7.

Bellinger, E. G. (1979) The response of algal populations to changes in lake water quality. In James, A. and Evison, L. (eds), *Biological indicators of water quality* pp. 9.1–9.27, Wiley, Chichester.

Benarde, M. A. (1970) *Our precarious habitat*, Norton, New York.

Benson-Evans, K. and Williams, P. F. (1975) Algae and Bryophytes. In Curds, C. R. and Hawkes, H. A., *Ecological aspects of used-water treatment* pp. 153–202, Academic Press, London.

Bergerson, E. P. and Galat, D. L. (1975) Coniferous tree bark: a lightweight substitute for limestone rock in barbeque basket macroinvertebrate samplers, *Water Res.*, **9,** 729–31.

Besch, W. K., Loseries, H. G., Meyer-Waarden, K. and Schmitz, W. (1974) Warntest zum Nachweis akut toxischer Konzentration von Wasserinhaltsstoffen, *Arch. Hydrobiol.*, **74,** 551–65.

Best, G. A. and Ross, S. L. (1977) *River pollution studies*, Liverpool University Press, Liverpool.

Bick, H. (1968) Autökologische und saprobiologische Untersuchungen an Susswasserciliaten, *Hydrobiologia*, **31,** 17–36.

Biró, P. (1979) Acute effects of the sodium salt of 2, 4-D on the early developmental stages of bleak, *Alburnus alburnus*, *J. Fish. Biol.*, **14,** 101–9.

Bitton, G. (1978) Survival of enteric viruses. In Mitchell, R. (ed), *Water pollution microbiology, vol. 2* pp. 273–99, Wiley, New York.

Blaylock, B. G. and Frank, M. L. (1979) A comparison of the toxicity of nickel to the developing eggs and larvae of carp (*Cyprinus carpio*), *Bull. Environ. Contam. Toxicol.*, **21,** 604–11.

Bonde, G. J. (1977) Bacterial indication of water pollution, *Adv. aquat. microbiol.*, **1,** 273–364.

Brinkhurst, R. O. (1970) Distribution and abundance of tubificid (Oligochaeta) species in Toronto Harbour, Lake Ontario, *J. Fish. Res. Bd. Can.*, **27,** 1961–9.

Brooker, M. P. and Edwards, R. W. (1973) Effects of the herbicide paraquat on the ecology of a reservoir. 1. Botanical and chemical aspects, *Freshwat. Biol.*, **3,** 157–76.

Brooker, M. P. and Edwards, R. W. (1974) Effects of the herbicide paraquat on the ecology of a reservoir III. Fauna and general discussion, *Freshwat. Biol.*, **4,** 311–35.

Brooker, M. P. and Edwards, R. W. (1975) Aquatic herbicides and the control of water weeds, *Water Res.*, **9,** 1–15.

Brooks, J. L. (1969) Eutrophication and changes in the composition of the zooplankton. In *Eutrophication: Causes, Consequences, Correctives*, pp. 236–55, National Academy of Sciences, Washington.

Brooks, R. R. and Rumsby, M. G. (1965) The biogeochemistry of trace element uptake by some New Zealand bivalves, *Limnol. Oceanogr.*, **10,** 521–7.

Brown, B. E. (1977) Uptake of copper and lead by a metal tolerant isopod *Asellus meridianus* Rac., *Freshwat. Biol.*, **7,** 235–44.

Brown, B. E. (1978) Lead detoxification by a copper-tolerant isopod, *Nature, Lond.*, **276,** 388–90.

Brown, V. M. (1976) Advances in testing the toxicity of substances to fish, *Chemy. Ind.*, **21,** 143–9.

Brown, V. M., Jordan, D. H. M. and Tiller, B. A. (1967) The effect of temperature on the acute toxicity of phenol to rainbow trout in hard water, *Water Res.*, **1,** 587–94.

Brown, V. M., Shurben, D. G. and Shaw, D. (1970) Studies on water quality and the absence of fish from some polluted English rivers, *Water Res.*, **4,** 363–82.

Bruce, A. M. (ed) (1979) *Utilization of sewage sludge on land – papers and proceedings*, Water Research Centre, Medmenham.

Brungs, W. A. (1969) Chronic toxicity of zinc to the fathead minnow (*Pimephales promelas* Rafinesque), *Trans. Am. Fish. Soc.*, **98,** 272–9.

Bryce, D., Caffoor, I. M., Dale, C. R. and Jarrett, A. F. (1978) *Macro-invertebrates and the bioassay of water quality: a report based on a survey of the River Lee*, Nelpress, London.

Bryson, R. A. and Murray, T. J. (1977) *Climates of hunger*, University of Wisconsin Press, Madison.

Buckley, E. N., Jonas, R. B. and Pfaender, F. K. (1976) Characterization of microbial isolates from the estuarine ecosystem: relationship of hydrocarbon utilization to ambient hydrocarbon concentrations, *App. Environ. Microbiol.*, **32,** 232–7.

Burnet, F. M. and White, D. O. (1972) *Natural history of infectious disease*, 4th edn, Cambridge University Press, London.

Butcher, R. W. (1932) Contribution to our knowledge of the ecology of sewage fungus, *Trans. Brit. mycol. Soc.*, **17,** 112–23.

Butcher, R. W. (1947) Studies on the ecology of rivers, VII. The algae of organically enriched waters, *J. Ecol.*, **35,** 186–91.

Cairns, J., Albaugh, D. W. and Busey, F. and Chancy, M. D. (1968) The sequential comparison index – a simplified method for non-biologists to estimate relative differences in biological diversity in stream pollution studies, *J. Wat. Pollut. Control Fed.*, **40,** 1607–13.

Cairns, J. and Dickson, K. L. (1971) A simple method for the biological assessment of the effects of waste discharges on aquatic bottom-dwelling organisms, *J. Wat. Pollut. Control Fed.*, **43,** 755–72.

Cairns, J., Dickson, K. L. and Westlake, G. F. (1974) Continuous biological monitoring to establish parameters for water pollution control, *7th Int. Conf. Wat. Pollut. Res.*, Paris 1974.

Cairns, J., Heath, A. G. and Parker, B. C. (1975) The effects of temperature upon the toxicity of chemicals to aquatic organisms, *Hydrobiologia*, **47,** 135–71.

Calamari, D. and Marchetti, R. (1973) The toxicity of mixtures of metals and surfactants to rainbow trout (*Salmo gairdneri* Rich.), *Water Res.*, **7,** 1453–64.

Capuzzo, J. M. (1979) The effects of halogen toxicants on survival, feeding and egg production of the rotifer *Brachionus plicatilis, Est. Coast. Mar. Sci.*, **8,** 307–16.

Carlson, R. W. and Drummond, R. A. (1978) Fish cough response – a method for evaluating quality of treated complex effluents, *Water Res.*, **12,** 1–6.

Carson, R. (1962) *Silent Spring*, Houghton-Mifflin, Boston.

Carter, C. E. (1978) The fauna of the muddy sediments of Lough Neagh, with particular reference to eutrophication, *Freshwat. Biol.*, **8,** 547–59.

Cember, H., Curtis, E. H. and Blaylock B. G. (1978) Mercury bioconcentration in fish: temperature and concentration effects, *Environ. Pollut.*, **17,** 311–9.

Chandler, J. R. (1970) A biological approach to water quality management, *Wat. Pollut. Control.*, **69,** 415–22.

Chanin, P. R. F. and Jefferies, D. J. (1978) The decline of the otter *Lutra lutra* L. in Britain: an analysis of hunting records and discussion of causes, *Biol. J. Linn. Soc.*, **10,** 305–28.

Cherry, D. S., Guthrie, R. K. and Harvey, R. S. (1974) Temperature influence on bacterial populations in three aquatic systems, *Water Res.*, **8,** 149–55.

Christensen, G., Hunt, E. and Fianot, J. (1977) The effect of methylmercuric chloride, cadmium chloride and lead nitrate on six biochemical factors of the brook trout (*Salvelinus fontinalis*), *Toxicol. appl. Pharm.,* **42,** 523–30.

Chutter, F. M. (1972) An empirical biotic index of the quality of water in South African streams and rivers, *Water Res.*, **66,** 19–30.

Chutter, F. M. and Noble, R. G. (1966) The reliability of a method of sampling stream invertebrates, *Arch. Hydrobiol.*, **62,** 95–103.

Clifford, H. T. and Stephenson, W. (1975) *An introduction to numerical classification*, Academic Press, San Francisco.

Coker, E. G. (1966) The value of liquid disgested sludge. 1. The effect of liquid sewage sludge on growth and composition of grass-clover swards in south-east England, *J. Agric. Sci.*, **67**, 91–7.

Collingwood, R. W. (1977) A survey of eutrophication in Britain and its effects on water supplies, *Technical Report no. 40*, Water Research Centre, Medmenham.

Colwell, R. R. and Sayler, G. S. (1978) Microbial degradation of industrial chemicals. In Mitchel, R. (ed), *Water pollution microbiology*, *vol. 2*, pp. 111–34, Wiley-Interscience, New York.

Convery, J. J. (1970) Treatment techniques for removing phosphorus from municipal wastewaters, *Water Pollution Control Res. Ser. No. 17010-01/70*, Washington.

Cook, S. E. (1976) Quest for an index of community structure sensitive to water pollution, *Environ. Pollut.*, **11,** 269–88.

Cooke, G. W. and Williams, R. J. B. (1973) Significance of man-made sources of phosphorus: fertilizers and farming, *Water Res.*, **7,** 19–33.

Coutant, C. C. (1962) The effect of heated water effluent upon the macro-invertebrate fauna of the Delaware river, *Proc. Penn. Acad. Sci.*, **36,** 58–71.

Coutant, C. C., Cox, D. K. and Moore, D. K. W. (1976) Further studies of cold-shock effects on susceptibility of young channel catfish to predation. In Esch, G. W. and McFarlane, R. W., *Thermal Ecology II* pp. 154–8, Technical Information Service, Springfield, Virginia.

Cover, E. C. and Harrel, R. C. (1978) Sequences of colonization, diversity, biomass, and productivity of macroinvertebrates on artificial substrates in a freshwater canal, *Hydrobiologia*, **59,** 81–95.

Crossman, J. S. and Cairns, J. (1974) A comparative study between two different artificial substrate samplers and regular sampling techniques, *Hydrobiologia*, **44,** 517–22.

Crowther, R. F. and Harkness, N. (1975) Anaerobic bacteria. In Curds, C. R. and Hawkes, H. A., *Ecological aspects of used-water treatment* pp. 65–91, Academic Press, London.

Curds, C. R. (1975) Protozoa. In Curds, C. R. and Hawkes, H. A., *Ecological aspects of used-water treatment* pp. 203–68. Academic Press, London.

Curds, C. R., Cockburn, A. and Vandyke, J. M. (1968) An experimental study of the role of the ciliated protozoa in the activated sludge process, *Wat. Pollut. Control*, **67,** 312–29.

Curtis, E. J. C. and Curds, C. R. (1971) Sewage fungus in rivers in the United Kingdom: the slime community and its constituent organisms, *Water Res.*, **5,** 1147–59.

Curtis, E. J. C., Delves-Broughton, J. and Harrington, D. W. (1971) Sewage fungus: studies of *Sphaerotilus* slimes using laboratory recirculating channels, *Water Res.*, **5,** 267–79.

Dafni, Z., Ulitzur, S. and Shilo, M. (1972) Influence of light and phosphate on toxin production and growth of *Prymnesium parvum*, *J. gen. Microbiol.*, **70,** 199–207.

Dakin, W. J. and Latarche, M. (1913) The plankton of Lough Neagh, *Proc. Roy. Ir. Acad.* (*B*), **30,** 20–96.

Dangerfield, B. J. (ed) (1979) *The structure and management of the British water industry*, Institution of Water Engineers and Scientists, London.

Danil'chenko, O. P. (1977) The sensitivity of fish embryos to the effect of toxicants, *J. Ichthyol.*, **17,** 455–63.

Dausend, K. (1931) Über die Atmung der Tubificiden, *Z. vergl. Physiol.*, **14,** 557–608.

Davis, J. C. (1978) Disruption of precopulatory behavior in the amphipod *Anisogammarus pugettensis* upon exposure to bleached kraft pulpmill effluent, *Water Res.*, **12,** 273–5.

Dawson, F. H. (1978) Aquatic plant management in semi-natural streams: the role of marginal vegetation, *J. Environ. Management*, **6,** 213–21.

Department of the Environment (1972) *Analysis of raw, potable and waste waters*, HMSO, London.

Department of the Environment (1973) *Report of a survey of the discharges of foul sewage to the coastal waters of England and Wales*, HMSO, London.

Department of the Environment (1978) *River pollution survey of England and Wales, updated 1975*, HMSO, London.

Deufel, J. (1972) Die Bakterien- und Keimzahlen im Oberlauf der Donau bis Ulm, *Arch. Hydrobiol. Suppl.*, **44,** 1–9.

Devey, D. G. and Harkness, N. (1973) The significance of man-made sources of phosphorus: detergents and sewage, *Water Res.*, **7,** 33–54.

Dillon, P. J. and Kirchner, W. D. (1975) The effects of geology and land use on the export of phosphorus from watersheds, *Water Res.*, **9,** 135–48.

Dillon, P. J. and Rigler, F. H. (1974) The phosphorus-chlorophyll relationship in lakes, *Limnol. Oceanogr.*, **19,** 767–73.

Dillon, P. J. and Rigler, F. H. (1975) A simple method for predicting the capacity of a lake for development based on trophic status, *J. Fish. Res. Bd. Canada*, **32,** 1519–31.

Downing, J. A. (1979) Aggregation, transformation, and the design of benthos sampling programs, *J. Fish Res. Bd. Canada*, **36,** 1454–63.

Doxat, J. (1977) *The living Thames. The restoration of a great tidal river*, Hutchinson Benham, London.

Dugan, R. P. (1972) *Biochemical ecology of water pollution*, Plenum Press, New York.

Dunst, R. C. (1974) Survey of lake rehabilitation techniques and experiences, *Tech. Bull. no. 75*, Department of Natural Resources, University of Wisconsin.

Dutka, B. J. (1973) Coliforms are an inadequate index of water quality, *J. Environ. Health*, **36,** 39–46.

Eden, G. E. (1979) Biological oxidation by percolating filters and other fixed-film devices. In *Water pollution control technology*, pp. 48–59. HMSO, London.

Edmondson, W. T. (1969) Eutrophication in North America. In *Eutrophication: Causes, Consequences, Correctives* pp. 124–49, National Academy of Sciences, Washington.

Edmondson, W. T. (1970) Phosphorus, nitrogen and algae in Lake Washington after diversion of sewage, *Science*, **169,** 690–91.

Edmondson, W. T. (1971) Phytoplankton and nutrients in Lake Washington. In Likens, G. (ed), *Nutrients and eutrophication: the limiting nutrient controversy*, American Society of Limnology and Oceanography 1, 172–93.

Edmondson, W. T. (1972a) The present condition of Lake Washington, *Verh. internat. Verein. Limnol.*, **18,** 284–91.

Edmondson, W. T. (1972b) Lake Washington. In Goldman, C. R. (ed), *Environmental quality and water development* pp. 281–98, Freeman and Co., New York.

Edmondson, W. T. (1974) Book review, *Limnol. Oceanog.*, **19,** 369–75.

Edmondson, W. T. (1979) Lake Washington and the predictability of limnological events, *Ergebn. Limnol.*, **13,** 234–41.

Edmondson, W. T., Anderson, G. C. and Peterson, D. R. (1956) Artificial eutrophication of Lake Washington, *Limnol. Oceanogr.*, **1,** 47–53.

Edwards, R. W. (1975) A strategy for the prediction and detection of effects of pollution on natural communities, *Schweiz. Z. Hydrol.* **37,** 135 43.

Edwards, R. W. and Brown, V. M. (1966) Pollution and fisheries: A progress report, *J. Inst. Water Pollut. Control, Lond.*, **66,** 63–78.

Edwards, R. W., Hughes, B. D. and Read, M. W. (1975) Biological survey in the detection and assessment of pollution. In Chadwick, M. J. and Goodman, G. T. (eds), *The ecology of resource degradation and renewal* pp. 139–56, Blackwell, Oxford.

Edwards, R. W. and Owens, M. (1965) The oxygen balance of streams. In Goodman, G. T., Edwards, R. W. and Lambert, J. M. (eds), *Ecology and the industrial society* pp. 149–72, Blackwell, Oxford.

Elliott, J. M. (1977) *Some methods for the statistical analysis of benthic invertebrates, Scientific Publication No. 25,* Freshwater Biological Association, Windermere.

Epstein, E. (1975) Effect of sewage sludge on some soil physical properties, *J. Environ. Qual.*, **4,** 139–42.

Esser, W. (1978) Über die Rolle sessiler Organismen auf die Selbstreinigungsgeschwindigkeit in Fliessgewassern, *Gas-u Wasserfach*, **119,** 582–86.

Eyres, J. P., Williams, N. V. and Pugh-Thomas, M. (1978) Ecological studies on Oligochaeta inhabiting depositing substrata in the Irwell, a polluted English river, *Freshwat. Biol.*, **8,** 25–32.

Fager, E. W. (1957) Determination and analysis of recurrent groups, *Ecology*, **38,** 586–95.

Fahy, E. (1975) Quantitative aspects of the distribution of invertebrates in the benthos of a small stream system in western Ireland, *Freshwat. Biol.*, **5,** 167–82.

Farmer, G. J., Ashfield, D. and Samant, H. S. (1979) Effects of zinc on juvenile Atlantic salmon *Salmo salar*: acute toxicity, food intake, growth and bio-accumulation, *Environ. Pollut.*, **19,** 103–17.

Fjerdingstad, E. (1965) Taxonomy and saprobic valency of benthic phytomicro-organisms, *Int. Rev. ges. Hydrobiol.*, **50,** 475–604.

Flannagan, J. F. (1970) Efficiencies of various grabs and corers in sampling freshwater benthos, *J. Fish. Res. Bd. Can.*, **27,** 1691–700.

Foster, R. B. and Bates, J. M. (1978) Use of mussels to monitor point source industrial discharges, *Environ. Sci. Technol.,* **12,** 958–62.

Fowler, D. L. and Mahan, J. N. (1975) *The pesticide review,* U.S. Dept. Agr., Agr. Stab. Cons. Serv., Washington, D.C.

Fox, H. M. (1954) Oxygen and haem in invertebrates, *Nature*, **174,** 355.

Frazier, J. M. (1979) Bioaccumulation of cadmium in marine organisms, *Environmental Health Perspectives*, **28,** 75–9.

Gaddum, J. H. (1948) *Pharmacology* 3rd edn, Oxford University Press, London.

Gameson, A. L. H. and Wheatland, A. B. (1958) The ultimate oxygen demand and course of oxidation of sewage effluents, *J. Inst. Sew. Purif., pt. 2*, 106–19.

Gaudy, A. F. (1972) Biochemical oxygen demand, In Mitchell, R. (ed), *Water pollution microbiology* pp. 305–32, Wiley Interscience, New York.

Geldreich, E. E. (1972) Water-borne pathogens. In Mitchell, R. (ed), *Water pollution microbiology*, pp. 207–41, Wiley-Interscience, New York.

Geldreich, E. E. and Kenner, B. K. (1969) Concepts of faecal streptococci in stream pollution, *J. Wat. Pollut. Control. Fed.*, **41,** R336–R352.

Gelin, C. (1978) The restoration of freshwater ecosystems in Sweden. In Holdgate, M. W. and Woodman, M. J. (eds), *The breakdown and restoration of ecosystems*, pp. 332–36, Plenum, New York.

Gerba, C. P., Goyal, S. M. and Melnick, J. L. (1977) Distribution of viral and bacterial pathogens in a coastal community, *Mar. Pollut. Bull.*, **8,** 279–82.

Ghetti, P. F. and Bonazzi, G. (1977) A comparison between various criteria for the interpretation of biological data in the analysis of the quality of running water, *Water Res.*, **11,** 819–31.

Goldsmith, F. B. and Harrison, C. M. (1976) Description and analysis of vegetation. In Chapman, S. B. (ed), *Methods in plant ecology*, pp. 85–155, Blackwell Scientific Publications, Oxford.

Goodland, R. (1977) Panamanian development and the global environment, *Oikos*, **29,** 195–208.

Gore, P. H., Wilson, S. and Capener, H. R. (1975) A sociological approach to the problem of water pollution, *Growth and Change*, **6,** 17–22.

Gower, A. M. and Buckland, P. J. (1978) Water quality and the occurrence of *Chironomus riparius* Meigen (Diptera: Chironomidae) in a stream receiving sewage effluent, *Freshwat. Biol.*, **8,** 153–64.

Gray, J. (1976) Are marine base-line surveys worthwhile?, *New Scientist*, **81,** 219–21.

Green, R. H. (1979) *Sampling design and statistical methods for environmental biologists*, Wiley, New York.

Hall, D. O. (1979) Solar energy conversion through biology, *Biologist*, **26,** 67–74.

Hamilton, A. L., Burton, W. and Flannagan, J. (1970) A multiple corer for sampling profundal benthos, *J. Fish. Res. Bd. Can.*, **27,** 1867–69.

Hammerton, D. (1972) The Nile River – a case history. In Oglesby, R. T., Carlson, C. A. and McCann, J. A. (ed), *River ecology and man*, pp. 171–214, Academic Press, New York.

Harding, J. P. C. and Whitton, B. A. (1978) Zinc, cadmium and lead in water, sediments and submerged plants of the Derwent Reservoir, Northern England, *Water Res.*, **12,** 307–16.

Hargreaves, J. W., Lloyd, E. J. K. and Whitton, B. A. (1975) Chemistry and vegetation of highly acidic streams, *Freshwat. Biol.*, **5,** 563–76.

Hargreaves, J. W., Mason, C. F. and Pomfret, J. R. (1979) A simplified biotic index for the assessment of biologically-oxidizable pollution in flowing waters, *Wat. Pollut. Control*, **78,** 98–105.

Hartmann, J. (1977) Fischereiliche Veranderungen in kulturbedingt eutrophierenden Seen, *Schweiz. Z. Hydrol.*, **39,** 243–54.

Haslam, S. M. (1978) *River plants*, Cambridge University Press, Cambridge.

Hatch, T. (1962) Changing objectives in occupational health, *Am. ind. Hyg. Ass. J.*, **23,** 1–7.

Hawkes, H. A. (1962) Biological aspects of river pollution. In Klein, L., *River pollution II. Causes and effects,* pp. 311–432, Butterworths, London.

Hawkes, H. A. (1963) *The ecology of waste water treatment*, Pergamon Press, Oxford.

Hawkes, H. A. (1975) River zonation and classification. In Whitton, B. (ed), *River ecology*, pp. 312–374, Blackwell Scientific Publications, Oxford.

Hawkes, H. A. (1978) River bed animals tell-tales of pollution. In Hughes, J. G. and Hawkes, H. A. (collators) *Biosurveillance of river water quality* pp. 55–77. Proceedings of Section K of the British Association for the Advancement of Science, Aston 1977.

Hawkes, H. A. (1979) Invertebrates as indicators of river water quality. In Jones, A. and Evison, L., *Biological indicators of water quality* pp. 2.1–2.49, John Wiley & Sons, Chichester.

Hawkes, H. A. and Davies, L. J. (1971) Some effects of organic enrichment on benthic invertebrate communities in stream riffles. In Duffey, E. A. and Watt, A. S. (eds), *The scientific management of animal and plant communities for conservation* pp. 271–93, Blackwell, Oxford.

Heit, M. and Fingerman, M. (1977) The influences of size, sex and temperature on the toxicity of mercury to two species of crayfishes, *Bull. Environ. Contam. Toxicol.*, **18,** 572–80.

Hellawell, J. (1977) Biological surveillance and water quality monitoring. In Alabaster, J. S. (ed), *Biological monitoring of inland fisheries* pp. 69–88, Applied Science Publishers Ltd., London.

Hellawell, J. M. (1978) *Biological surveillance of rivers*, Water Research Centre, Medmenham.

Herbert, D. W. M. (1961) Freshwater fisheries and pollution control, *Proc. Soc. Wat. Treat. J.*, **10,** 135–56.

Hermanutz, R. O. (1978) Endrin and malathion toxicity to flagfish (*Jordanella floridae*), *Arch Environ. Contam. Toxicol.*, **7,** 159–68.

Hester, F. E. and Dendy, J. S. (1962) A multiple plate sampler for aquatic macroinvertebrates, *Trans. Am. Fish. Soc.*, **91,** 420–21.

Higgins, I. J. and Burns, R. G. (1975) *The chemistry and microbiology of pollution*, Academic Press, London.

Hill, W. F., Akin, E. W. and Benton, W. H. (1971) Detection of viruses in water: a review of methods and application, *Water Res.*, **5,** 967–95.

Hillbricht-Ilkowska, A. and Zdanowski, B. (1978) Effect of thermal effluents and retention time on lake functioning and ecological efficiencies in plankton communities, *Int. Revue ges. Hydrobiol.*, **63,** 609–17.

Hocutt, C. H., Kaesler, R. L., Masnik, M. T. and Cairns, J. (1974) Biological assessment of water quality in a large river system: an evaluation of a method for fishes, *Arch. Hydrobiol.* **74,** 448–62.

Holdgate, M. W. (1979) *A perspective of environmental pollution*, Cambridge University Press, Cambridge.

Holdway, P. A., Watson, R. A. and Moss, B. (1978) Aspects of the ecology of *Prymnesium parvum* (Haptophyta) and water chemistry in the Norfolk Broads, England, *Freshwat. Biol.*, **8,** 295–311.

Holmes, N. T., Whitton, B. A. and Hargreaves, J. W. (1978) A coded list of freshwater macrophytes of the British Isles, *WDU Water Archive Manual Series*, **4,** 1–201.

Holt, R. F., Timmons, D. R. and Latterell, J. L. (1970) Accumulation of phosphates in water, *J. agric. Fd. Chem.*, **18,** 781–4.

Honda, T. and Finkelstein, R. A. (1979) Selection and characteristics of a *Vibrio cholerae* mutant lacking the A (ADP-ribosylating) portion of the cholera enterotoxin, *Proc. Natl. Acad. Sci. U.S.A.*, **76,** 2052–56.

Howarth, R. S. and Sprague, J. B. (1978) Copper lethality to rainbow trout in waters of various hardness and pH, *Water Res.*, **12,** 455–62.

Huet, M. (1954) Biologie, profils en long et en travers des eaux courantes, *Bull. fr. Piscic.*, **175,** 41–53.

Hughes, B. D. (1975) A comparison of four samplers for benthic macro-invertebrates inhabiting coarse river deposits, *Water Res.*, **9,** 61–9.

Hughes, B. D. (1978) The influence of factors other than pollution on the value of Shannon's diversity index for benthic macro-invertebrates in streams, *Water Res.*, **12,** 357–64.

Hughes, G. M. (1964) Fish respiratory homeostasis. In *Homeostasis and feedback mechanisms, Symp. Soc. exp. Biol.*, **18** pp. 81–107, Academic Press, London.

Hughes, G. M. (1976) Polluted fish respiratory physiology. In Lockwood, A. P. M. (ed) *Effects of pollutants on aquatic organisms*, pp. 163–83, Cambridge University Press, Cambridge.

Hunter, J. B. (1978) The role of the toxicity test in water pollution control, *Wat. Pollut. Control*, **77,** 384–94.

Hurlbert, S. H. (1969) A coefficient of interspecific association, *Ecology*, **50,** 1–9.

Hynes, H. B. N. (1960) *The biology of polluted waters*, Liverpool University Press, Liverpool.

Hynes, H. B. N. (1970) *The ecology of running waters*, Liverpool University Press, Liverpool.

Imboden, D. M. and Gächter, R. (1978) A dynamic lake model for trophic state prediction, *Ecol. Modelling*, **4,** 77–98.

Inverson, W. P. and Brinckman, F. E. (1978) Microbial metabolism of heavy metals, In Mitchell, R., *Water pollution microbiology vol. 2* pp. 201–32, Wiley-Interscience, New York.

Jaccard, P. (1908) Nouvells recherches sur la distribution florale, *Bull. Soc. Vaud. Sci. Nat.*, **44,** 223–270.

Jago, P. H. (1977) Review of filtration techniques for tertiary treatment of sewage effluents, *W.R.C. Technical Report*, **64,** 1–16.

Jeffers, J. N. R. (1978) *Design of experiments. Statistical Checklist 1* Natural Environment Research Council, London.

Jeffers, J. N. R. (1979) *Sampling. Statistical Checklist 2*, Natural Environment Research Council, London.

Jenkins, S. H. (1970) Biological filtration. In *Water pollution control engineering* pp. 46–50, HMSO, London.

Jeris, J. S. and Owens, R. W. (1978) Biological fluidized beds for nitrogen control. In Wanielista, M. P. and Eckenfelder, W. W. (eds), *Advances in water and wastewater treatment. Biological nutrient removal*, pp. 199–213, Ann Arbor Science, Ann Arbor, Michigan.

Jewell, W. J., Morris, G. R., Price, D. R., Gunkel, W. W., Williams, D. W. and Loehr, R. C. (1974) Methane generation from agricultural wastes: a review of concepts and future applications, *Paper No. NA74–107, Amer. Soc. Ag. Engns.*, St. Joseph.

Johnson, W. E. and Vallentyne, J. R. (1971) Rationale, background and development of experimental lake studies in northwestern Ontario, *J. Fish. Res. Bd. Can.*, **28,** 123–28.

Jónasson, P. M. (1955) The efficiency of sieving techniques for sampling freshwater bottom fauna, *Oikos,* **6,** 183–207.

Jones, H. R. and Peters, J. C. (1977) Physical and biological typing of unpolluted rivers, *W.R.C. Technical Report 41*, 1–48.

Jones, J. G. (1979) *A guide to methods for estimating microbial numbers and biomass in fresh water. Scientific publication no. 39*, Freshwater Biological Association, Windermere.

Jones, J. R. E. (1947) The oxygen consumption of *Gasterosteus aculeatus* L. in toxic solutions, *J. exp. Biol.*, **23,** 298–311.

Jorgensen, S. E. (1980) *Lake management:* Pergamon Press, Oxford. lake model, *Ecol. Modelling*, **4,** 253–78.

Jørgensen, S. E. (1980) *Lake management:* Pergamon Press, Oxford.

Jusatz, H. (1977) Cholera. In Howe, G. M., *A world geography of human diseases* pp. 131–43, Academic Press, London.

Kaesler, R. L. and Cairns, J. (1972) Cluster analysis of data from limnological surveys of the upper Potomac river, *Am. Midl. Nat.*, **88,** 56–67.

Kaesler, R. L., Carins, J. and Bates, J. M. (1971) Cluster analysis of non-insect macro-invertebrates of the Upper Potomac River, *Hydrobiologia*, **37,** 173–81.

Kalinin, G. P. and Shiklomanov, J. A. (1974) *World water balance and water resources of the Earth*, USSR National Committee for the International Hydrological Decade, Leningrad.

Källqvist, T. and Meadows, B. S. (1978) The toxic effect of copper on algae and rotifers from a soda lake (Lake Nakuru, East Africa), *Water Res.*, **12,** 771–5.

Ketchum, B. H. (1972) *The water's edge: critical problems of the coastal zone*, MIT Press, Cambridge, Mass.

Klein, L. (1962) *River pollution II. Causes and effects*, Butterworths, London.

Klein, L. (1966) *River pollution III. Control*, Butterworths, London.

Klekowski, E. J. (1978) Screening aquatic ecosystems for mutagens with fern bioassays, *Environ. Hlth. Persp.*, **27,** 99–102.

Kolkwitz, R. and Marsson, M. (1980) Ökologie der pflanzlichen Saprobien, *Ber. Dt. Botan. Ges.*, **261,** 505–19.

Kolkwitz, R. and Marsson, M. (1909) Ökologie der tierischen Saprobien, *Int. Rev. ges. Hydrobiol.*, **2,** 125–52.

Korte, F. (1974) Global inputs and burdens of chemical residues in the biosphere: the problem and control measures. In *Comparative studies of food and environmental contamination* pp. 3–22, International Atomic Energy Agency, Vienna.

Koryak, M., Shapiro, M. A. and Sykora, J. L. (1972) Riffle zoobenthos in streams receiving acid mine drainage, *Water Res.*, **6,** 1239–47.

Krause, A. (1977) On the effect of marginal tree-rows with respect to the management of small lowland streams. *Aquat. Bot.*, **3,** 185–92.

Krebs, C. T. and Valiela, I. (1978) Effect of experimentally applied

chlorinated hydrocarbons on the biomass of the fiddler crab, *Uca pugnax* (Smith), *Est. Coastal Mar. Sci.*, **6,** 375–86.
Krebs, C. T., Valiela, I., Harvey, G. R. and Teal, J. M. (1974) Reduction of field populations of fiddler crabs by uptake of chlorinated hydrocarbons, *Mar. Pollut. Bull.*, **5,** 140–2.
Lack, T. J. and Lund, J. W. G. (1974) Observations and experiments on the phytoplankton of Blelham Tarn, English Lake District. 1. The experimental tubes, *Freshwat. Biol.*, **4,** 399–415.
Langford, T. E. (1970) The temperature of a British river upstream and downstream of a heated discharge from a power station, *Hydrobiologia*, **35,** 353–75.
Langford, T. E. (1972) A comparative assessment of thermal effects in some British and North American rivers. In Oglesby. R. T., Carlson, C. A. and McCann, J. A., *River ecology and man* pp. 319–51, Academic Press, New York.
Langford, T. E. (1975) The emergence of insects from a British river, warmed by power station cooling water. Part II. The emergence patterns of some species of Ephemeroptera, Trichoptera and Megaloptera in relation to water temperature and river flow, upstream and downstream of the cooling water outfalls, *Hydrobiologia*, **47,** 91–133.
Langford, T. E. and Aston, R. J. (1972) The ecology of some British rivers in relation to warm water discharges from power stations, *Proc. R. Soc. Lond. B*, **180,** 407–19.
Langford, T. E. and Howells, G. (1977) The use of biological monitoring in the freshwater environment by the electrical industry in the UK. In Alabaster, J. S. (ed), *Biological monitoring of inland fisheries* pp. 115–24, Applied Science Publishers, London.
Larrick, S. R., Dickson, K. L., Cherry, D. S. and Cairns, J. (1978) Determining fish avoidance of polluted water, *Hydrobiologia*, **61,** 257–65.
Learner, M. A. and Edwards, R. W. (1963) The toxicity of some substances to *Nais* (Oligochaeta), *Proc. Soc. Wat. Treat. Exam.*, **12,** 161–8.
Learner, M. A., Lochhead, G. and Hughes, B. D. (1978) A review of the biology of British Naididae (Oligochaeta) with emphasis on the lotic environment, *Freshwat. Biol.*, **8,** 357–75.
Learner, M. A., Williams, R., Harcup, M. and Hughes, B. D. (1971) A survey of the macro-fauna of the River Cynon, a polluted tributary of the River Taff (South Wales), *Freshwat. Biol.*, **1,** 339–67.
Lee, G. F., Rast, W. and Jones, R. A. (1978) Eutrophication of

water bodies: Insights for an age-old problem, *Environ. Sci. Technol.*, **12,** 900–8.
Leeming, J. B. (1978) The role of biology and its practical application to river management in England and Wales. In *Elaboration of the scientific bases for monitoring the quality of surface water by hydrobiological indicators. Pollution report no. 3* pp. 39–45 Department of the Environment, London.
Leeuwangh, P. (1978) Toxicity tests with Daphnids: its application in the management of water quality, *Hydrobiologia*, **59,** 145–8.
Letterman, R. D. and Mitsch, W. J. (1978) Impact of mine drainage on a mountain stream in Pennsylvania, *Environ. Pollut.*, **17,** 53–73.
Likens, G. E., Bormann, F. H., Johnson, N. M., Fisher, D. W. and Pierce, R. S. (1970) Effects of forest cutting and herbicide treatment on nutrient budgets in the Hubbard Brook Watershed ecosystem, *Ecol. monogr.*, **40,** 23–47.
Lindh, G. (1979) Water and food production. In Biswas, M. R. and Biswas, A. K. (eds) *Food, climate and man* pp. 52–72. Wiley, New York.
Lingaraja, T., Sasi Bhushana Rao, P. and Venugopalan, V. K. (1979) DDT induced ethological changes in estuarine fish, *Env. Biol. Fish.*, **4,** 83–8.
Lloyd, R. (1960) The toxicity of zinc sulphate to rainbow trout, *Ann. appl. Biol.*, **48,** 84–94.
Lloyd, R. (1961) Effect of dissolved oxygen concentrations on the toxicity of several poisons to rainbow trout (*Salmo gairdneri*, Richardson), *J. exp. Biol.*, **38,** 447–55.
Lloyd, R. (1972) Problems in determining water quality criteria for freshwater fisheries, *Proc. R. Soc. Lond. B*, **180,** 439–49.
Lloyd, R. and Herbert, D. W. M. (1962) The effect of the environment on the toxicity of poisons to fish, *Instn. publ. Hlth. Engrs.*, **61,** 132–45.
Lloyd, R. and Orr, L. D. (1969) The diuretic response by rainbow trout to sublethal concentrations of ammonia, *Water Res.*, **3,** 335–44.
Lloyd, R. and Swift, D. J. (1976) Some physiological responses by freshwater fish to low dissolved oxygen, high carbon dioxide, ammonia and phenol with particular reference to water balance. In Lockwood, A. P. M., *Effects of pollutants on aquatic organisms* pp. 47–69, Cambridge University Press, Cambridge.
Loehr, R. C., Martin, C. S. and Rast, W. (eds.) (1980) *Phosphorus management strategies for lakes*, Ann Arbor Science, Michigan.

LRE (1979) Flash! Instrument replaces fish, *Environ. Sc. Technol.*, **13,** 646.

Lund, J. W. G. (1959) Biological tests on the fertility of an English reservoir water (Stocks Reservoir, Bowland Forest), *J. Inst. Wat. Eng.*, **13,** 527–49.

Lund, J. W. G. (1975) The uses of large experimental tubes in lakes. In *The effects of storage on water quality*, Water Research Centre, Medmenham, pp. 291–312.

Lund, J. W. G. (1978) Experiments with lake phytoplankton in large enclosures, *46th Annual Report of the Freshwater Biological Association* pp. 32–9.

Lundgren, D. G., Vestal, J. R. and Tabita, F. R. (1972) The microbiology of mine drainage pollution. In Mitchell, R. (ed) *Water pollution microbiology*, pp. 69–88, Wiley-Interscience, New York.

Lvovich, M. I. (1973) The global water balance, *EOS 54*, no. 1.

Macan, T. T. (1958) Methods of sampling the bottom fauna in stony streams, *Mitt. int. Verein. theor. angew. Limnol.*, **8,** 1–21.

Macan, T. T. (1959) *A guide to freshwater invertebrate animals*, Longman, London.

Macan, T. T. and Kitching, A. (1972) Some experiments with artificial substrata, *Verh. Internat. Verein. Limnol.*, **18,** 213–30.

Macdonald, S. M. and Mason, C. F. (1976) The status of the otter (*Lutra lutra* L.) in Norfolk, *Biological Conservation*, **9,** 119–24.

Macdonald, S. M. and Mason, C. F. (1980) The demise of the otter, *Biologist*, **27,** 140–2.

Macdonald, S. M., Mason, C. F. and Coghill, I. S. (1978) The otter and its conservation in the River Teme catchment, *J. appl. Ecol.*, **15**, 373–84.

Mackay, D. W., Tayler, W. K. and Henderson, A. R. (1978) The recovery of the polluted Clyde Estuary, *Proc. Roy. Soc. Edinb.*, **76B,** 135–52.

Maitland, P. S. (1977) *A coded checklist of animals occurring in freshwater in the British Isles*, Institute of Terrestrial Ecology, Edinburgh.

Maitland, P. S. (1978) *Biology of fresh waters*, Blackie, Glasgow and London.

Maitland, P. S. and Morris, K. H. (1978) A multi-purpose modular limnological sampler, *Hydrobiologia*, **59,** 187–95.

Martin, E. A. (ed) (1976) *A dictionary of life sciences*, Macmillan, London.

Maruoka, S. (1978) Estimation of toxicity using cultured mammalian cells of the organic pollutants recovered from Lake Biwa, *Water Res.*, **12,** 371–5.

Mason, C. F. (1977a) Populations and production of benthic animals in two contrasting shallow lakes in Norfolk, *J. anim. Ecol.*, **46,** 147–72.

Mason, C. F. (1977b) The performance of a diversity index in describing the zoobenthos of two lakes, *J. appl. Ecol.*, **14,** 363–7.

Mason, C. F. (1978) Artificial oases in a lacustrine desert, *Oecologia* (*Berl.*), **36,** 93–102.

Mason, C. F. and Bryant, R. J. (1974) The structure and diversity of the animal communities in a broadland reedswamp, *J. Zool.*, **172,** 289–302.

Mason, C. F. and Bryant, R. J. (1975) Changes in the ecology of the Norfolk Broads, *Freshwat. Biol.*, **5,** 257–70.

Mason, C. F. and Macdonald, S. M. (1976) Aspects of the breeding biology of the snipe, *Bird Study*, **23,** 33–8.

Mason, W. T., Anderson, J. B. and Morrison, G. E. (1967) A limestone-filled artificial substrate sampler-float unit for collecting macroinvertebrates in large streams, *Prog. Fish. Cult.*, **29,** 74.

Mason, W. T., Weber, C. I., Lewis, P. A. and Julian, E. C. (1973) Factors affecting the performance of basket and multiplate macroinvertebrate samplers, *Freshwat. Biol.*, **3,** 409–36.

Matson, E. A., Hornor, S. G. and Buck, J. D. (1978) Pollution indicators and other micro-organisms in river sediment, *J. Wat. Pollut. Control. Fed.*, **50,** 13–9.

Mayes, R. A. McIntosh, A. W. and Anderson. V. L. (1977) Uptake of cadmium and lead by a rooted aquatic macrophyte (*Elodea canadensis*), *Ecology*, **58**, 1176–80.

McEwen, F. L. and Stephenson, G. R. (1979) *The use and significance of pesticides in the environment*, John Wiley and Sons, New York.

McFarlane, R. W. (1976) Fish diversity in adjacent ambient, thermal and past thermal freshwater streams. In Esch, G. W. and McFarlane, R. W. *Thermal Ecology II* pp. 268–71, Technical Information Centre, Springfield, Virginia.

McFarlane, R. W., Moore, B. C. and Williams, S. E. (1976) Thermal tolerance of stream cyprinid minnows. In Esch, G. W. and McFarlane, R. W. *Thermal Ecology II* pp. 141–4, Technical Information Centre, Springfield, Virginia.

McKim, J. M., Eaton, J. G. and Holcombe, G. W. (1978) Metal toxicity to embryos and larvae of eight species of freshwater fish–II: Copper, *Bull. Environm. Contam. Toxicol.*, **19,** 608–16.

McLean, R. O. and Jones, A. K. (1975) Studies of tolerance to heavy metals in the flora of the rivers Ystwyth and Clarach, Wales, *Freshwat. Biol.*, **5,** 431–44.

McLeay, D. J. and Brown, D. A. (1975) The effects of acute exposure to bleached kraft pulp mill effluent on carbohydrate metabolism of juvenile coho salmon (*Oncorhynchus kisutch*) during rest and exercise, *J. Fish. Res. Bd. Can.*, **32,** 753–60.

MacLeod, J. C. and Smith, L. L. (1966) Effect of pulpwood fiber on oxygen consumption and swimming endurance of the fathead minnow, *Pimephales promelas, Trans. Am. Fish. Soc.*, **95,** 71–84.

Meier, P. G., Penrose, D. L. and Polak, L. (1979) The rate of colonization by macro-invertebrates on artificial substrate samplers, *Freshwat. Biol.*, **9,** 381–92.

Metcalf, T. G. (1978) Indicators for viruses in natural waters. In Mitchell, R. (ed) *Water pollution microbiology. vol. 2*, pp. 301–24, Wiley, New York.

Middlebrooks, E. J., Falkenborg, D. H. and Maloney, T. E. (1974) *Modelling the eutrophication process*, Ann Arbor, Michigan.

Morgan, W. S. G. (1978) The use of fish as a biological sensor for toxic compounds in potable water, *Prog. Wat. Tech.*, **10,** 395–8.

Morgan, W. S. G. (1979) Fish locomotor behavior patterns as a monitoring tool, *J. Wat. Pollut. Control. Fed.*, **51,** 580–9.

Morgan, W. S. G. and Kühn, P. C. (1974) A method to monitor the effects of toxicants upon breathing rate of largemouth bass (*Micropterus salmoides*) Lacépède), Water Res., **8,** 67–77.

Moriarty, F. (1975a) The dispersal and persistence of p, p^1-DDT. In Chadwick, M. J. and Goodman, G. T., *The ecology of resource degradation and renewal*, pp. 31–47, Blackwell Scientific Publications, Oxford.

Moriarty, F. (1975b) Exposures and residues. In Moriarty, F. (ed) *Organochlorine insecticides: persistent organic pollutants* pp. 29–72, Academic Press, London.

Moss, B. (1972) Studies on Gull Lake, Michigan. II. Eutrophication–evidence and prognosis, *Freshwat. Biol.*, **2,** 309–20.

Moss, B. (1978) The ecological history of a mediaeval man-made lake, Hickling Broad, Norfolk, United Kingdom, *Hydrobiologia*, **60,** 23–32.

Mossewitch, N. A. (1961) Sauerstoffdefizit in den Flussen des Westsibirischen Tieflandes, seine Ursachen und Einflüsse auf die aquatische Fauna, *Verh. int. Verein. theor. angew. Limnol.*, **14,** 447–50.

Mount, D. I. (1968) Chronic toxicity of copper to fathead minnows (*Pimephales promelas*, Rafinesque), *Water Res.*, **2,** 215–33.

Mount, D. I. and Stephan, C. E. (1967) A method of establishing acceptable toxicant limits for fish – malathion and the butoxyethanol ester of 2, 4-D, *Trans. Am. Fish. Soc.*, **96,** 185–93.

Muirhead-Thomson, R. C. (1971) *Pesticides and freshwater fauna*, Academic Press, London.

Muirhead-Thomson, R. C. (1978a) Relative susceptibility of stream macro-invertebrates to temephos and chlorpyrifos, determined in laboratory continuous-flow systems, *Arch. Environm. Contam. Toxicol.*, **7,** 129–37.

Muirhead-Thomson, R. C. (1978a) Lethal and behavioral impact of chlorpyrifos methyl and temephos on select stream macroinvertebrates: experimental studies on downstream drift, *Arch. Environm. Contam. Toxicol.*, **7,** 139–47.

Mulder, E. G. and Veen, W. L. van (1963) Investigations on the *Sphaerotilus – heptothrix* group, *Antonie van Leeuwenhoek J. Microbiol. Serol 29*, 121–53.

Murphy, P. M. (1978) The temporal variability in biotic indices, *Environ. Pollut.*, **17,** 227–36.

Myers, N. (1979) *The sinking ark*, Pergamon Press, London.

National Water Council (1978) *Water Industry Review,* NWC, London.

Natural Environment Research Council (1971) *National Angling Survey 1970*, NERC, London.

Natural Environment Research Council (1977) Ecological research on seabirds, *NERC Publ. Series C, no. 18*, 1–48.

Nature Conservancy Council (1977) *Otters 1977*. First report of the Joint NCC/SPNC Otter Group, 26 pp., London.

Naumann, E. (1919) Nägra synpunkter angående planktons ökologi. Med Sarskild hänsyntill fytoplankton, *Svensk. bot. Tidskr.*, **13,** 129–58.

Nelson, D. J., Kaye, S. V. and Booth, R. S. (1972) Radionuclides in river systems. In Oglesby, R. T., Carlson, C. A. and McCann, J. A., *River ecology and man*, pp. 367–87, Academic Press, New York.

Nemerow, N. L. (1974) *Scientific stream pollution analysis*, McGraw-Hill, New York.

Neumann, D. (1961) Der Einfluss des Eisenangebotes auf die Hamoglobinsynthese und die Entwicklung der *Chironomus*-Larve, *Z. Naturforsch.* **16b,** 820–4.

Newbold, C. (1975) Herbicides in aquatic systems, *Biol. Conserv.*, **7,** 97–118.

Newbold, C. (1977) Wetlands and agriculture. In Davidson, J. and

Lloyd, R., *Conservation and agriculture* pp. 59–79, Wiley, Chichester.

Nuttall, P. M. and Purves, J. B. (1974) Numerical indices applied to the results of a survey of the macro-invertebrate fauna of the Tamar catchment (south-west England), *Freshwat. Biol*, **4,** 213–22.

Oden, B. J. (1979) The freshwater littoral meiofauna in a South Carolina reservoir receiving thermal effluents, *Freshwater Biol.*, **9,** 291–304.

Okun, D. A. (1977) *Regionalization of water management*, Applied Sciences Publishers, Barking, Essex.

Oswald, W. J. (1977) Determinants of feasiblity in bioconversion of solar energy. In Castellani, A. (ed). *Research in photobiology* pp. 371–86, Plenum Press, London.

Owens, M. (1970) Nutrient balance in rivers, *Wat. Treat. Exam.*, **19,** 239–47.

Palmer, C. M. (1959) Algae in water supplies, *U.S. Public Health Serv. Publ. No. 657.*

Palmer, M. F. (1968) Aspects of the respiratory physiology of *Tubifex tubifex* in relation to its ecology, *J. Zool.*, **154,** 463–73.

Palmer, M. F. and Chapman, G. (1970) The state of oxygenation of haemoglobin in the blood of living *Tubifex* (Annelida), *J. Zool.*, **161,** 203–9.

Pantle, R. and Buck, H. (1955) Die biologische Uberwachung der Gewasser und die Darstellung der Ergebnisse, *Gas-u Wasserfach*, **96,** 604.

Pascoe, D. and Beattie, J. H. (1979) Resistance to cadmium by pretreated rainbow trout alevins, *J. Fish. Biol.*, **14,** 303–8.

Patrick, R. (1954) Diatoms as an indication of river change, *Proc. 9th Industr. Waste Conf., Purdue Univ. Engng. Extn. Ser.*, **87,** 325–30.

Payne, A. G. (1975) Responses of the three test algae of the Algal Assay Procedure: Bottle Test, *Water Res.*, **9,** 437–45.

Pearson, J. M. (1978) Computing and data processing services, *Water Data Unit Tech. Mem. 1 (Revised)*, 1–11.

Pentelow, F. T. K., Butcher, R. W. and Grindley, J. (1938) An investigation of the effects of milk wastes on the Bristol Avon, *Fish. Invest., Lond.*, **1, 4, 1,**

Pereira, H. C. (1973) *Land use and water resources in temperate and tropical climates*, Cambridge University Press, Cambridge.

Perkins, E. J. (1979) The effects of marine discharges on the ecology of coastal waters. In James, A. and Evinson, L. (eds) *Biological indicators of water quality* pp. 12-1–12-42, Wiley, Chichester.

Phillips, D. J. H. (1977) The use of biological indicator organisms to

monitor trace metal pollution in marine and estuarine environments – a review, *Environ. Pollut.*, **13,** 281–317.
Phillips, G. L., Eminson, D. and Moss, B. (1978) A mechanism to account for macrophyte decline in progressively eutrophicated freshwaters, *Aquat. Bot.*, **4,** 103–26.
Pickering, Q. H. (1968) Some effects of dissolved-oxygen concentrations upon the toxicity of zinc to the bluegill, *Lepomis macrochirus*, Raf., *Water Res.*, **2,** 187–94.
Pike, E. B. (1975) Aerobic Bacteria. In Curds, C. R. and Hawkes, H. A., *Ecological aspects of used-water treatment*, pp. 1–64, Academic Press, London.
Pike, E. B. (1978) The design of percolating filters and rotary biological contactors, including details of international practice, *W.R.C. Technical Report 93*, 1–44.
Pimental, D. and Goodman, N. (1978) Ecological basis for the management of insect populations, *Oikos*, **30,** 422–37.
Pitcairn, C. E. R. and Hawkes, H. A. (1973) The role of phosphorus in the growth of *Cladophora*, *Water Res.*, **7,** 159–71.
Poels, C. L. M., Snoek, O. I. and Huizenga, L. J. (1978) Toxic substances in the Rhine River, *Ambio*, **7,** 218–25.
Poldoski, J. E. (1979) Cadmium bioaccumulation assays. Their relationship to various ionic equilibria in Lake Superior water, *Environ. Sci. Technol.*, **13,** 701–6.
Porter, E. (1978) *Water management in England and Wales*, Cambridge University Press, Cambridge.
Porter, K. S. (1975) *Nitrogen and phosphorus, food production, waste and the environment*, Ann Arbor Science, Michigan.
Potten, A. H. (1972) Maturation ponds. Experiences in their operation in the United Kingdom as a tertiary treatment process for a high quality sewage effluent, *Water Res.*, **6,** 373–91.
Preston, A. (1972) Artificial radio-activity in freshwater and estuarine systems, *Proc. R. Soc. Lond. B.*, **180,** 421–36.
Preston, A. (1974) Application of critical path analysis techniques to the assessment of environmental capacity and the control of environmental waste disposal. In *Comparative studies of food and environmental contamination* pp. 573–83 International Atomic Energy Agency, Vienna.
Price, D. H. A. (1970) Sewage treatment. In *Water pollution control engineering* pp. 34–45, HMSO, London.
Price, D. R. H. (1978) Fish as indicators of water quality, *Wat. Pollut. Control*, **77,** 285–96.
Price, D. R. H. (1979) Fish as indicators of water quality. In James,

A. and Evinson, L. (eds) *Biological indicators of water quality* pp. 8.1–8.23, Wiley, Chichester.

Price, D. R. H. and Pearson, M. J. (1979) The derivation of quality conditions for effluents discharged to freshwaters, *Wat. Pollut. Control*, **78,** 118–38.

Prins, R. and Black, W. (1971) Synthetic webbing as an effective macrobenthos sampling substrate in reservoirs, *Spec. Publs. Am. Fish. Soc.*, **8,** 203–8.

Rathner, M. and Sonneborn, M. (1979) Biologisch Wirksame Östrogene in Trink und Abwasser, *Forum Städte-Hygiene*, **30,** 45–9.

Rathore, H. S., Sanghvi, P. K. and Swarup, H. (1979) Toxicity of cadmium chloride and lead nitrate to *Chironomus tentans* larvae, *Environ. Pollut.*, **18,** 173–7.

Ray, S. N. and White, W. J. (1979) *Equisetum arvense* – an aquatic vascular plant as a biological monitor for heavy metal pollution, *Chemosphere*, **3,** 125–8.

Reish, D. (1973) The use of marine benthic animals in monitoring the marine environment, *J. Environ. Plann. Pollut. Control.* **1,** 32–8.

Resh, V. H. and Unzicker, J. D. (1975) Water quality monitoring and aquatic organisms: the importance of species identification, *J. Wat. Pollut. Control. Fed.*, **47,** 9–19.

Rodda, J. C., Sheckley, A. C. and Tan, P. (1978) Water resources and climatic change, *J. Inst. W. Eng. and Sci.* **32,** 76–83.

Rodhe, W. (1969) Crystallization of eutrophication concepts in northern Europe. In *Eutrophication: Causes, Consequences, Correctives*, pp. 50–64, National Academy of Sciences, Washington.

Ryding, S.-O. and Forsberg, C. (1976) Six polluted lakes: a preliminary evaluation of the treatment and recovery processes, *Ambio*, **5,** 151–6.

Särkkä, J. (1979) Mercury and chlorinated hydrocarbon in zoobenthos of Lake Päijänne, Finland, *Arch. Environm. Contam. Toxicol.*, **8,** 161–73.

Särkkä, J., Hattula, M. L., Passivirta, J. and Janatuinen, J. (1978) Mercury and chlorinated hydrocarbons in the food chain of Lake Päijänne, Finland, *Holarctic Ecol.*, **1,** 326–32.

Sastry, K. V. and Gupta, P. K. (1978) Alterations in the activity of some digestive enzymes of *Channa punctatus*, exposed to lead nitrate, *Bull. Environm. Contam. Toxicol.*, **19,** 549–55.

Saunders, P. J. W. (1976) *The estimation of pollution damage*, Manchester University Press, Manchester.

Sauter, S., Buxton, K. S., Macek, K. J. and Petrocelli, S R. (1976)

Effects of exposure to heavy metals on selected freshwater fish. Toxicity of copper, cadmium, chromium, and lead to eggs and fry of seven fish species, *Ecological Research Series,* EPA-600/3-76-105.

Say, P. J., Diaz, B. M. and Whitton, B. A. (1977) Influence of zinc on Lotic plants. 1. Tolerance of *Hormidium* species to zinc, *Freshwat. Biol.*, **7,** 357–76.

Say. P. J. and Whitton, B. A. (1977) Influence of zinc on Lotic plants. II. Environmental effects on toxicity of zinc to *Hormidium rivulare*, *Freshwat. Biol.*, **7,** 377–884.

Scavia, D. and Robertson, A. (eds) (1979) *Perspectives on lake ecosystem modelling*. Ann Arbor Science, Michigan.

Scharf, B. W. (1979) A fish test alarm device for the continual recording of acute toxic substances in water, *Arch. Hydrobiol.*, **85,** 250–6.

Schindler, D. W. and Fee, E. J. (1973) Diurnal variation of dissolved inorganic carbon and its use in estimating primary production and CO_2 invasion in Lake 227, *J. Fish. Res. Bd. Can.*, **30,** 1501–10.

Schindler, D. W., Fee, E. J. and Ruszczynski, T. (1978) Phosphorus input and its consequences for phytoplankton standing crop and production in the Experimental Lakes Area and in similar lakes, *J. Fish. Res. Bd. Can.*, **35,** 190–6.

Schindler, D. W., Kling, H., Schmidt, R. V., Prokopowich, J., Frost, V. E., Reid, R. A., and Capel, M. (1973) Eutrophication of Lake 227 by addition of phosphate and nitrate: the second, third and fourth years of enrichment, 1970, 1971, and 1972, *J. Fish. Res. Bd. Can.*, **30,** 1415–40.

Schindler, J. E. (1971) Food quality and zooplankton nutrition, *J. anim. Ecol.*, **40,** 589–96.

Sellers, C. M., Heath, A. G. and Bass, M. L. (1975) The effect of sublethal concentrations of copper and zinc on ventilatory activity, blood O_2 and pH in rainbow trout (*Salmo gairdneri*), *Water Res.*, **6,** 217–30.

Sewell, W. R. and Barr, L. R. (1978) Water administration in England and Wales. Impacts of reorganization, *Water Res. Bull.*, **14,** 337–48.

Shuval, H. I. (1975) The case for microbial standards fór bathing beaches. In Gameson, A. L. H. (ed), *Discharge of sewage from sea outfalls*, Pergamon Press, London.

Shuval, H. I. and Katzenelson, E. (1972) The detection of enteric viruses in the water environment. In Mitchell, R. (ed) *Water pollution microbiology* pp. 347–61., Wiley, New York.

Skidmore, J. F. (1970) Respiration and osmo-regulation in rainbow trout with gills damaged by zinc sulphate, *J. exp. Biol.*, **52,** 484–94.
Slack, J. G. (1977) River water quality in Essex during and after the 1976 drought, *Effl. Wat. Treatment J.*, **17,** 575–8.
Sládeček, V. (1979) Continental systems for the assessment of river water quality. In James, A. and Evison, L. (eds) *Biological indicators of water quality*, pp. 3.1–3.32. Wiley, Chichester.
Sloey, W. E., Spangler, F. L. and Fetter, C. W. (1978) Management of freshwater wetlands for nutrient assimilation. In Good, R. E., Whigham, D. F. and Simpson, R. L., *Freshwater wetlands* pp. 321–40, Academic Press, New York.
Smart, G. R. (1978) Investigations of the toxic mechanisms of ammonia to fish-gas exchange in rainbow trout (*Salmo gairdneri*) exposed to acute lethal concentrations, *J. Fish. Biol.*, **12,** 93–104.
Smith, G. R. (1976) Botulism in waterfowl, *Wildfowl*, **27,** 129–38.
Smith, M. J. and Heath, A. G. (1979) Acute toxicity of copper, chromate, zinc and cyanide to freshwater fish: effect of different temperatures, *Bull. Environm. Contam. Toxicol.*, **22,** 113–19.
Smith, R. V. (1977) Domestic and agricultural contributions to the inputs of phosphorus and nitrogen to Lough Neagh, *Water Res.*, **11,** 453–9.
Smith, W. E. and Smith, A. M. (1975) *Minamata*, Holt, Rinehart and Winston, New York.
Solbé, J. F. de L. G. (1971) Aspects of the biology of the lumbricids *Eiseniella tetraedra* (Savigny) and *Dendrobaena rubida* (Savigny) *f. subrubicunda* (Eisen) in a percolating filter, *J. appl. Ecol.*, **8,** 845–67.
Solbé, J. F. de L. G. (1973) The relation between water quality and the status of fish populations in Willow Brook, *Water Treat. Examin.*, **22,** 41–61.
Solbé, J. F. de L. G. (1977) Water quality, fish and invertebrates in a zinc-polluted stream. In Alabaster, J. S. (ed) *Biological monitoring of inland fisheries* pp. 97–105, Applied Science Publishers Ltd., London.
Solbé, J. F. de L. G. and Cooper, V. A. (1976) Studies on the toxicity of copper sulphate to stone loach *Noemacheilus barbatulus* (L.) in hard water, *Water Res.*, **10,** 523–7.
Solbé, J. F. de L. G. and Tozer, J. S. (1971) Aspects of the biology of *Psychoda alternata* (Say.) and *P. severini parthenogenetica* Tonn. (Diptera) in a percolating filter, *J. appl. Ecol.*, **8,** 835–44.
Solbé, J. F. de L. G., Williams, N. V. and Roberts, H. (1967) The colonization of a percolating filter by invertebrates, and their effect

on settlement of humus solids, *Wat. Pollut. Control.*, **66,** 423–48.

Søensen, T. (1948) A method of establishing groups of equal amplitude in plant sociology based on similarity of species content and its application to analyses of the vegetation on Danish commons, *Biol. Skr. (k. danske vidensk. Selsk N.S.)*, **5,** 1–34.

Southgate, V. R., Wijk, H. B. van and Wright, C. A. (1976) Schistosomiasis at Loum, Cameroun; *Schistosoma haematobium*, *S. intercalatum* and their natural hybrid, *Z. Parasitenk*, **49,** 145–59.

Southworth, G. R., Beauchamp, J. J. and Schmieders, P. K. (1978) Bioaccumulation potential of polycyclic aromatic hydrocarbons in *Daphnia pulex*, *Water Res.*, **12,** 973–77.

Spigarelli, S. A. and Smith, D. W. (1976) Growth of salmonid fishes from heated and unheated areas of Lake Michigan- measured by RNA-DNA ratios. In Esch, G. W. and McFarlane, R. W. *Thermal Ecology II* pp. 100–105 Technical Information Centre, Springfield, Virginia.

Sprague, J. B. (1964) Lethal concentrations of copper and zinc for young Atlantic salmon, *J. Fish. Res. Bd. Can.*, **21,** 17–26.

Sprague, J. B. (1969) Measurement of pollutant toxicity to fish. I. Bioassay methods for acute toxicity, *Water Res.*, **3,** 793–821.

Sprague, J. B. (1970) Measurement of pollutant toxicity to fish. II. Utilizing and applying bio-assay results, *Water Res.*, **4,** 3–32.

Sprague, J. B. (1971) Measurement of pollution toxicity to fish. III. Sublethal effects and safe concentrations. *Water Res.*, **5,** 245–66.

Statton, G. W. and Corke, C. T. (1979) The effect of cadmium ion on the growth, photosynthesis, and nitrogenase activity of *Anabaena inaequalis*, *Chemosphere*, **5,** 277–82.

Sterling, C. (1971) Aswan Dam looses a flood of problems, *Life*, **70** **(5)** 46–7.

Stokes, J. L. (1954) Studies on the filamentous sheathed iron bacterium *Sphaerotilus natans*, *J. Bacteriol.*, **67,** 278–91.

Stoner, A. W. and Livingston, R. J. (1978) Respiration, growth and food conversion efficiency of pinfish (*Lagodon rhomboides*) exposed to sublethal concentrations of bleached kraft mill effluent, *Environ Pollut.*, **17,** 207–17.

Stones, T. (1979) A critical examination of the uses of the BOD test, *Effl, Wat. Treat. J.*, **19,** 252–4.

Stott, B. and Cross, D. G. (1973) The reactions of roach (*Rutilus rutilus* (L.)) to changes in the concentration of dissolved oxygen and free carbon dioxide in a laboratory channel, *Water Res.*, **7,** 793–805.

Streeter, H. and Phelps, E. (1925) A study of the purification of the Ohio River, *U. S. Publ. Health Service Bulletin No. 146*, Washington, D.C.

Sugden, B. and Lloyd, L. (1950) The clearing of turbid waters by means of the ciliate *Carchesium*–a demonstration, *J. Inst. Sew. Purif.*, **1,** 16–20.

Sugiura, K., Washino, T., Hattori, M., Sato, E. and Goto, M. (1978) Accumulation of organochlorine compounds in fishes. Differences of accumulation factors by fishes, *Chemosphere*, **9,** 359–64.

Surber, E. W. (1937) Rainbow trout and bottom fauna production in one mile of stream, *Trans. Am. Fish. Soc.*, **66,** 193–202.

Svedberg. T. and Eriksson-Quensel, I. B. (1934) The molecular weight of erythrocruorin, *J. Amer. Chem. Soc.*, **56,** 1700–6.

Szczesny, B. (1974) The effect of sewage from the town of Krynica on the benthic invertebrate communities of the Kryniczanka stream, *Acta. Hydrobiol.*, **16,** 1–29.

Teppen, T. C. and Gammon, J. R. (1976) Distribution and abundance of fish populations, in the Middle Wabash River. In Esch. G. W. and McFarlane, R. W., *Thermal Ecology II* pp. 272–83, Technical Information Centre, Springfield, Virginia.

Tevlin, M. P. (1978) An improved experimental medium for freshwater toxicity studies using *Daphnia magna*, *Water Res.*, **12,** 1017–24.

Thomas, E. A. (1969) The process of eutrophication in central European lakes. In *Eutrophication: Causes, Consequences, Correctives*, pp. 29–49, National Academy of Sciences, Washington.

Thomas, W. A. (1976) Attitudes of professionals in water management towards the use of water quality indices, *J. Environ. Management,* **4,** 325–38.

Tomlinson, T. E. (1971) Nutrient losses from agricultural land, *Outlook on Agriculture*, **6,** 272–8.

Tomlinson, T. G. and Williams, I. L. (1975) Fungi. In Curds, C. R. and Hawkes, H. A., *Ecological aspects of used-water treatment*, pp. 93–152, Academic Press, London.

Toms, R. G. (1975) Management of river water quality. In Whitton, B. A. (ed), *River Ecology* pp. 538–64, Blackwell Scientific Publications, Oxford.

UNESCO/WHO (1978) *Water quality surveys. A guide for the collection and interpretation of water quality data*, UNESCO/WHO, Paris.

United States Environmental Protection Agency (1971) *Algal Assay Procedure: bottle test*, Nat. Eutroph. Res. Prog., Corvallis, Oregon, USA.

Van Donsel, D. J. and Geldreich, E. E. (1971) Relationships of Salmonellae to faecal coliforms in bottom sediments, *Water Res.*, **5,** 1075–87.

Veen, W. L. van, Mulder, E. G. and Deinema, M. H. (1978) The *Sphaerotilus-Leptothrix* group of bacteria, *Microbiol. Rev.* **42,** 329–56.

Venosa, A. D. (1975) Lysis of *Sphaerotilus natans* swarm cells by *Bdellovibrio* bacteriovirus, Appl. Microbiol. **29,** 702–5.

Verma, S. R., Tyagi, A. K., Pal, N. and Dalela, R. C. (1979) *In vivo* effects of the syndets Idet^R 5L and Swanic^R 6L on the ATPase activity in the teleost, *Channa punctatus, Arch. Environm. Contam. Toxicol.*, **8,** 241–6.

Vollenweider, R. A. (1968) *Scientific fundamentals of the eutrophication of lakes and flowing waters, with particular reference to nitrogen and phosphorus as factors in eutrophication*, OECD, Paris, 192 pp.

Vollenweider, R. A. (1969) Möglichkeiten und Grenzen elementarer Modelle der Stoffbilanz von Seen, *Arch. Hydrobiol.*, **66,** 1–36.

Vollenweider, R. A. (1975) Input-output models with special reference to the phosphorus loading concept in limnology, *Schweiz. Z. Hydrol.*, **37,** 53–84.

Vollenweider, R. A. and Dillon, P. J. (1974) The application of phosphorus loading concept to eutrophication research, *N.R.C. Tech. Rep. 13690*, 42 pp.

Walsh, F. (1978) Biological control of mine drainage. In Mitchell, R., *Water pollution microbiology, vol. 2* pp. 377–89, Wiley–Interscience, New York.

Walsh, F. and Mitchell, R. (1975) Mine drainage pollution reduction by inhibition of iron bacteria, *Water Res.*, **9,** 525–28.

Wanielista, M. P. and Eckenfelder, W. W. (eds) (1978) *Advances in water and wastewater treatment. Biological nutrient removal*, Ann Arbor Science, Michigan.

Ward, D. V., Howes, B. L. and Ludwig, D. F. (1976) Interactive effects of predation pressure and insecticide (Temephos) toxicity on populations of the marsh fiddler crab *Uca pugnax, Mar. Biol.*, **35,** 119–26.

Warner, R. E. (1967) Bio-assays for micro-chemical environmental contamination with special reference to water supplies, *Bull. World Health Org.*, **36,** 181–207.

Warren, C. E. (1971) *Biology and water pollution control*, W. B. Saunders and Co., Philadelphia.

Water Research Centre (1974) Some methods of utilizing organic waste waters, *Notes on water pollution*, **67,** 1–4.

Water Research Centre (1977) Pollution from urban run-off, *Notes on water research*, **12,** 1–4.

Webb, P. W. and Brett, J. R. (1972) The effects of sublethal concentrations of whole bleached kraft mill effluent on the growth and food conversion efficiency of underyearling sockeye salmon, *J. Fish. Res. Bd. Can.*, **29,** 1555–63.

Weber, C. A. (1907) Aufbau und Vegetation der Moore Norddeutschlands, *Beibl. Bot. Jahrb.,* **90,** 19–34.

Weglenska, T., and Hillbricht-Ilkowska, A. (1975) The effect of mineral fertilization on the structure and functioning of various trophic types of lakes. Part II. The effect of mineral fertilization on zoo-plankton, benthic fauna and tripton sedimentation, *Polskie Archw. Hdrobiol.*, **22,** 233–50.

Wert, F. S. and Henderson, U. B. (1978) Feed fish effluents and reel in savings, *Wat. Wastes Eng.*, **15,** 38–9; 43–4.

Westlake, D. F. and Edwards, R. W. (1957) Director's report, *Rep. Freshwat. biol. Ass. Brit. Emp.*, **25,** 35–7.

Westlake, G. F. and Van der Schalie, W. H. (1977) Evaluation of an automated biological monitoring system at an industrial site. In Cairns, J., Dickson, K. L. and Westlake, G. F. (eds). *Biological monitoring of water and effluent quality*, pp. 30–7, American Society for Testing and Materials, Philadelphia.

Wetzel, R. G. (1968) Dissolved organic matter and phytoplankton productivity in marl lakes, *Mitt. Int. Ver. Limol.*, **14,** 261–70.

White, J. C., Hammond, R. A., Wooding, N. H. and Brehmer, M. L. (1976) Temperature as a growth accelerator in the spot (Teleost: Scianidae). In Esch, G. W. and McFarlane, R. W., *Thermal Ecology II*, pp. 113–7, Technical Information Centre, Springfield, Virgina.

White, R. W. G. and Williams, W. P. (1978) Studies on the ecology of fish populations in the Rye Meads sewage effluent lagoons, *J. Fish Biol.*, **13,** 379–400.

Whitton, B. A. (1970) Biology of *Cladophora* in freshwaters, *Water Res.*, **4,** 457–76.

Whitton, B. A., Holmes, N. T. H. and Sinclair, C. (1978) A coded list of 1,000 freshwater algae of the British Isles, *WDU Water Archive Manual Series*, **2,** 1–335.

Wilhm, J. L. and Dorris, T. C. (1968) Biological parameters for water quality criteria, *BioScience*, **18,** 477–81.

Williams, N. V. and Taylor, H. M. (1968) The effects of *Psychoda alternata* (Say.) (Diptera) and *Lumbricillus rivalis* (Levinson) (Enchytraeidae) on the efficiency of sewage treatment in percolating filters, *Water Res.*, **2,** 139–50.

Williamson, M. H. (1972) *The analysis of biological populations*, Edward Arnold, London.

Wilson, R. S. and McGill, J. D. (1977) A new method of monitoring water quality in a stream receiving sewage effluent, using chironomid pupal exuviae, *Water Res.*, **11,** 959–62.

Wilson, R. S. and McGill, J. D. (1979) The use of chironomid pupal exuviae for biological surveillance of water quality, *Water Data Unit Tech. Mem. 18*, 1–20.

Windle-Taylor, E. (1978) The relationship between water quality and human health: medial aspects, *Royal Soc. Hlth. J.*, **98,** 121–9.

Winner, R. W., Van Dyke, J. S., Caris, N. and Farrel, M. P. (1975) Response of the macro-invertebrate fauna to a copper gradient in an experimentally-polluted stream, *Verh. Internat, Verein. Limnol.*, **19,** 2121–7.

Withers, G. K. (1978) Unregulated Potomac River causes supply problems in Washington, D.C., *Wat. Sew. Wks.*, **125** (9), 108–10.

Wollan, E., Davis, R. D. and Jenner, S. (1978) Effects of sewage sludge on seed germination, *Environ, Pollut.*, **17,** 195–205.

Wolters, W. R. and Coutant, C. C. (1976) The effect of cold shock on the vulnerability of young bluegill to predation. In Esch. G. W. and McFarlane, R. W., *Thermal Ecology II*, pp. 162–4, Technical Information Service, Springfield, Virginia.

Wolverton, B. C. and McDonald, R. C. (1978) Bioaccumulation and detection of trace levels of cadmium in aquatic systems by *Eichhornia crassipes, Environ. Health Perspectives*, **27,** 161–4.

Wood, G. (1975) *An assessment of eutrophication in Australian inland waters*, Australian Government Publishing Service, Canberra, 238 pp.

Wood, R. B. and Gibson, C. E. (1973) Eutrophication and Lough Neagh, *Water Res.*, **7,** 173–87.

Woodiwiss, F. S. (1964) The biological system of stream classification used by the Trent River Board, *Chemy. Ind.*, **11,** 443–7.

Wright, D. A. (1978) Heavy metal accumulation by aquatic invertebrates, *Applied Biology*, **3,** 331–94.

Zehnder, A. J. B. (1978) Ecology of methane formation. In Mitchell, R. (ed), *Water pollution microbiology, Vol. 2*, pp. 349–76, Wiley-Interscience, New York.

INDEX TO GENERA AND SPECIES

GENERAL INDEX